A Little Bit of Land

A Little Bit of Land

JESSICA GIGOT

Oregon State University Press   Corvallis

Library of Congress Cataloging-in-Publication Data

Names: Gigot, Jessica, author.
Title: A little bit of land / Jessica Gigot.
Description: Corvallis, OR : Oregon State University Press, 2022.
Identifiers: LCCN 2022027362 | ISBN 9780870712135 (paperback) | ISBN
   9780870712142 (ebook)
Subjects: LCSH: Gigot, Jessica. | Farm life—Washington (State)—Skagit
   County. | Women farmers—Washington (State)—Skagit County—
   Biography. | Farmers—Washington (State)—Skagit County—Biography. |
   Sustainable agriculture—Skagit County.
Classification: LCC S521.5.W2 G54 2022 | DDC 630.92 [B]—dc23/]
   eng/20220705
LC record available at https://lccn.loc.gov/2022027362

♾ This paper meets the requirements of ANSI/NISO Z39.48-1992
(Permanence of Paper).

First published in 2022 by Oregon State University Press
Second printing 2023
Printed in the United States of America

Oregon State University
OSU Press

Oregon State University Press
121 The Valley Library
Corvallis OR 97331-4501
541-737-3166 • fax 541-737-3170
www.osupress.oregonstate.edu

As we work to heal the earth, the earth heals us.
—Robin Wall Kimmerer

Virtue, like harmony, cannot exist alone;
a virtue must lead to harmony between one creature and another.
—Wendell Berry

for June and Eloise

A Note about Memoir

This book is a work of creative nonfiction. Some names have been changed and some haven't. All scenes, dialogue, and details are from my own memory and so they are admittedly biased. I have done my best to represent the people and places in my life accurately.

Wind

THE WIND HAS TAKEN OFF AGAIN. This is one of the hazards of living in the delta flats—streaks of insufferable wind. My old barn and the noble chestnut along the driveway have withstood these gusts for over eighty years, but there is no telling how the new greenhouse and duck coop will fare. I have spent many nights listening to sixty-mile-an-hour squalls whistle through the lath and plaster of this old farmhouse. Sometimes it all feels like an experiment, my life and this farm. I am trying to make a life and a living on this land.

The neighbors on three sides are potato fields. From the upstairs window of this small Craftsman where I sit and write, I can see the vacant potato rows from last fall's harvest. Migratory trumpeter swans waddle their way over and between ridges, claiming the last morsels of tuber. The view from this window offers an openness that is both nostalgic and desolate, especially in the muddy heart of winter. On a clear day, I can see some of Washington's gems, islands like Orcas and Lummi to the west and the luminous Mount Baker, an active volcano, to the east. The farm is 11.1 feet above sea level. The river is 1.5 miles away, and as the old-timers like to say, it overflows like a teacup.

I raise sheep and grow a wild mix of things, from basil to calendula flowers, all on ten acres. Most of the farms in this valley are male-owned and multigenerational, and they operate on several hundred acres—huge compared to mine, but small by national standards. Think wheat in the Palouse or corn and soybeans in the Midwest. It's all relative. Potatoes and berries are prominent crops in these fields, but over ninety types of crops grow well here, including cabbage seed, tulips, leeks, and pickling cucumbers. The Skagit Valley, where I live, is the last vestige of intact, continuous farmland along the I-5 corridor west of the Cascade Mountains.

⌁

Flooding and wind are the least of my worries. They are merely the recurring aches of living on a farm in a floodplain near sea level. In this northwestern part of Washington, I can feel grateful for fairly mild winters with minimal, sometimes nonexistent, snowfall. I rarely have to endure the nose-hair-freezing temperatures I experienced as a child sledding down my grandmother's cul-du-sac in Wisconsin or, as a young adult, navigating the icy sidewalks of my New England college campus. The mild, marine climate here is forgiving and generous and great for growing spuds. At least that is what the potato farmers say.

The turn from fall to winter is a quiet release on the farm. I can sit and reflect on the wind and other mysteries of this valley. Winter is mud season. That's all I see from this window and during my regular walks along the slough. Rain and river water leave patterns in the fields, lakes of water that remain for months after a hard rain and reflect the evening sky's pinks and blues. During this season, I'm reminded of Maxine Kumin's portrayal of the struggling seedling in her poem "Mud":

> Pods will startle apart,
> pellets be seized with a fever
> and as the dark gruel thickens,
> life will stick up a finger.

Inside winter's silent dormancy there's always the anticipation of change, of new life that will "stick up a finger" by spring. Winter allows me to read books and write my own poems in a state of luxurious denial about the summer mayhem and hard work up ahead. I do know that there is no true stopping point in winter, though, just a downshift into a slower pace of

doing and thinking and being. Growth and emergence are always imminent despite the apparent inertia of this time of year.

What does it mean to love a place? To know you are in the right place?

My educational trail from poetry to biology to poetry again has taught
me one thing—the Earth needs more companions, especially those who
are creative and willing to stay. As the great agrarian champion and poet
Wendell Berry writes, "If change is to come, it will have to come from the
margins." My story is one of hundreds of stories of aspiring small farmers
trying to reimagine the food system. In my adult life, I am becoming land
literate and learning about what is possible, internally and externally, when
a person stays put. In my heart, I feel the push and pull of gratitude and
sorrow daily—relief in finally feeling like I belong somewhere, along with
an acute awareness that, like all things in nature, change and transforma-
tion are inevitable.

This precious rest time will not last long. I am indoors, insulated from
the elements. Sitting at my desk, I sensuously grasp my pencil and slowly
sip my coffee. There is so much I still don't know. Over a decade on this
farm and I am still not used to the wind. It keeps distracting me. Tree
branches dance outside along the sheep pasture, and loose slivers of hay
escape the feeder and blow down the driveway. I know the gusts will die
down soon, possibly by midday.

When we have visitors on the farm, the question I am asked most fre-
quently is, "Did you grow up on a farm?" I usually pause, let out a terse
laugh, and reply, "Definitely not."

In my office, I stare at the blank page. It is a long story and I still don't
quite know how it ends. There is a lot I want to explain. How I got here,
why I wanted this, what I might lose, and, most importantly, how this little
bit of land has changed me.

The Eastside, 2001

IT'S SEPTEMBER 11 AND I am slowly waking up in my childhood bedroom, still decorated with lilac walls and the matching paisley bedspread I picked out in middle school. My dad just left for work, early as usual, but I hear the garage door below me reopening followed by the rapid clatter of footsteps. Then, he is yelling for my mom and me to come and watch the news with him. "This is unreal," he bellows from the bottom of the stairs. "I had to come back." The three of us sit on the couch and end up staying there all day. Our loyal springer spaniel, Sundae (named by my younger sister, who thought she looked like vanilla ice cream with chocolate sauce), is nestled in my lap. She has grayed and widened since I left for college four years ago. We all sit in silence.

⊷

Two towers engulfed in smoke, eventually imploding to dust and rubble. This would be one of my last memories of my parents together, and married, in our home.

⊷

That May I had graduated from college in Vermont and returned to Washington to work at a summer camp in the San Juan Islands. When camp ended, I had come home to figure out my next steps. We lived on "the eastside," a nickname for the many suburban towns east of Lake Washington outside of Seattle. I had been offered a nanny position with one of the Seattle families from camp. I'd also fallen deeply in love with one of my fellow counselors and was contemplating following him to his small southern Oregon town.

My sister had already left home. She'd finished high school and moved to Los Angeles that summer, more than ready to relocate and start her own adult life. I, on the other hand, was a twenty-two-year-old college graduate returning to the Pacific Northwest in my dented Volvo, desperate to avoid becoming part of what *Time* magazine had called the "culture of brokenness"—that is, the generation of placeless young people disillusioned by the government and absent a strong faith or home life to fall back on. I had successfully completed a degree in biology, but still felt somewhat rudderless.

⌁

My parents had met in junior high in Green Bay, Wisconsin. They dated throughout high school while going to separate Catholic schools, all girls and all boys. My father went on to college at Notre Dame University in Indiana and my mother went to Marquette University in Milwaukee. It was a three-hour trip, either by car or by bus, around the southern tip of Lake Michigan to see each other. In the summer of 1972, after graduating—my mother with a degree in English and my father with a degree in marketing—they got married and began working in Chicago. My mom was a copywriter for Sears and my dad got a job at a downtown advertising firm. Between them they had two degrees and that's all. There was no other choice but to make their own way.

Eventually they moved up to Madison, Wisconsin, working together at a smaller ad firm until I was born. After a scare at daycare when they thought I had meningitis and would need a spinal tap (it ended up being a bad ear infection), my mother decided to stay home with me. This division of labor made sense at the time. After watching her mother raise five kids, two younger sisters and two older brothers, it also seemed doable. My father would continue working as the sole breadwinner and she would take care of the house and kids for the next several decades. She did some freelance work for a few years, before throwing herself into PTA board positions.

Since I moved around a lot in my childhood, I never had a strong connection to any particular house, bedroom, or even state. The feeling of home had always been a distant one. People assumed my father worked in the military, not advertising. From my birthplace in Wisconsin, our family moved to Connecticut, then to California, before settling in the swollen suburban ring around Seattle when I was eleven. With every move, I felt more uprooted.

✦

Each year, my family traveled back to the Midwest to visit our relatives in Wisconsin. That was as close as I came to farmland—staring through a rental car window at the plowed fields, herds of Holsteins, and stoic silos. However, I remember feeling a deep sense of nostalgia as we drove through those cornfields and tilled pastures bleeding into housing developments and golf courses. Although my ancestors did not have a long-standing connection to the land (we were paper salesmen and pharmacists), agriculture and animal husbandry were so firmly rooted in the DNA of most small Wisconsin towns, I felt woven into the agrarian history.

My grandmother, a Wisconsin native, was my main teacher in the kitchen, showing me all the necessary, often arduous, recipe steps. She canned and baked most food from scratch. Since she was a chemistry major in college before getting married and having five children, she understood the science of baking, and she approached her culinary tasks with great discipline and focus.

Her chocolate chip cookies were a demonstration of scientific precision. Each ingredient perfectly handled and mixed, each step of the recipe followed with caution and care. She tended tomato, strawberry, and rhubarb plants in her garden and made the perfect pie crust, both buttery and flakey. Her special vinaigrette dressing, something she concocted using a wooden spoon to measure out the ingredients, seemed to me like pure alchemy until she shared the secret of her recipe with us in her last days.

"Jessie, grab my blue apron from the laundry room," she'd yell.

All her ingredients were perfectly arranged on the counter. My grandmother would patter back and forth within her small kitchen, muttering instructions to herself and me as I flipped through her long metal recipe box. "You have to get the exact amount of each ingredient or the cookies won't be the same" she said as she scraped off the crumbling edge of sticky brown sugar with a knife. With her hand mixer she beat the eggs and butter for no less than five minutes, even though they looked plenty combined to me. We'd spend this time together in the kitchen every time I visited; it was what I looked forward to most.

As a child growing up away from my grandmother's kitchen, my favorite foods tended toward Schwan's Tortellini,™ Hot Pockets,™ or Potato Buds™ (with a lot of butter). My mother tried to coordinate meals, but cooking was not her passion, and between our afterschool activities and my dad's work schedule, our family wasn't home together very often. I ate dinner in front of the TV most nights, and my most wholesome meals were on holidays. On those occasions there was a deliberate effort to make food, like pies, from scratch. Thanksgiving was my favorite meal not just because it was a holiday, but because my mother went all out with the traditional meal: red potatoes, green beans, turkey with homemade gravy, and a pecan pie. Most nights though, as time went on, we all learned to eat alone and at our own convenience.

As an awkward preteen, I often found myself wandering into the forest behind our house. The rich, earthy smells and luscious stands of ferns beckoned me. Entering a forest was like disappearing into another world that was both magical and slightly dangerous. Our neighborhood was less than three years old when we moved there, a product of the Microsoft-fueled housing boom. The houses and lawns on our street, which was a brand-new addition, glowed with neon newness while the bordering trees swayed and, sometimes at night, sobbed. Those widowed trees, left solitary from their dense, familiar stands, were prone to fall on houses. A combination of wet soil and the loss of a support root system made the trees very unstable. Most of our neighbors had the cedars behind their houses cut down before any damage was done.

When our house was quiet, which it often was, I'd scour my parent's collection of magazines and books for pictures of other places. For years I held on to a torn-out picture of a farming family: two parents, two kids, a cat, and a dog, all curled up in their quilt-covered bed on a Sunday morning. I took it from one of my mother's old *National Geographic* magazines on one of the neatly curated bookshelves in my dad's office. The picture was from an issue featuring New Zealand sheep farmers. I longed for the life I saw in that photo—the resiliency of it, perhaps—but I wasn't sure why. In my early teenage years, I already craved the deep domesticity and rooted-ness I imagined underscored these farmers' livelihood, and I admired the ruggedness of their existence, their direct connection to the land.

In high school, I worked at a landscaping company near our home. I had no experience, but they hired me anyway. That was my first outside job working with plants. At their nursery, my duties were to move inventory and keep all the potted plants alive over the long, hot summer days. Since I did not have experience in landscaping or gardening beyond helping my mom trim our front hedge, I elected to maintain the nursery floor. I made sure every plant on the lot was healthy and hydrated. It was repetitive work, not boring, but meditative. It took me a while to get the hang of it, to quiet my mind enough to focus.

Kip, the nursery owner, kept a vibrant stand of dahlias on an acre plot beyond the main office and nursery floor, and he sold cut flowers in the office. The dahlia patch was out by an old hammock near a small stream where I would often take my lunch breaks. Throughout the summer, as the dahlias came on in their brightness, I was shocked. I'd never been around so many blooms at once, and the gradation between tender peaches and browns, bold reds and yellows, was spectacular. Spending every day out-side that summer fulfilled a deep longing in me. When I was out in the dahlia patch, weary from a long day of lifting and watering, I'd find myself in a trance, staring into the center of each flower. As I absorbed the humid summer air, I would feel a calm come over my body, an exhausted cen-teredness I'd never before experienced.

⊷

When I entered college, it was with pleasant memories from my job at the nursery, a newfound affinity for plants, and an aspiration to study English and write poetry. I'd encountered Emily Dickinson in high school, and it was her work that fueled my initial interest in writing. I appreciated how Dickinson was humbled by the natural world and in awe of its mystery. Her view of nature was a holistic blend of reason and profound intuition:

> Nature is what we know—
> Yet have no art to say—
> So impotent Our Wisdom is
> To her Simplicity.

I wanted to explore that idea—that despite all that we, as humans, can learn through science and observation, our wisdom is impotent compared to the intricate workings of the natural world—by writing my own poems. However, after a few weeks in a freshman year ecology class with a badass female forest ecologist, I changed my mind.

Her sandy blonde hair was pulled back into a low ponytail. She didn't wear makeup, and for most classes was dressed in earth-toned pants and shirts. Her authority, her knowledge, wooed me. I was starstruck by her confidence as she entered the classroom and impressed by the photos of her most recent research trip deep within the forests of Alaska. I wanted to be her; I wanted to know what she knew about plants and trees and all their intimate secrets. Her laboratory exercises dressed us in waders looking for caddisfly larvae under river rocks. We trudged up mountains, past neon-clothed hunters, during the vibrant New England fall. I was hooked.

One of her first lectures was about Lake Washington—specifically, how it had died. I laughed when I first heard that in class. The lake of my summers, the one I jumped into regularly on hot summer nights from an abandoned freeway off-ramp with our gaggle of friends, was definitely not dead. (She was right. From 1941 to 1963, disposal of untreated sewage into the lake caused all the life below the surface to die off in a matter of decades. Only after several years of rehabilitation did it become alive again, and swimmable.)

I pocketed my poems and declared myself a biology major. What experience did I have to write about anyway? I wanted to understand the workings of the natural world and how to care for fragile landscapes thrown off balance by human acts of destruction and desecration. In my mind at the time, this was also the responsible career choice—to put down my pen and pick up the caliper, the calculator, the scale.

After I officially switched to science, I navigated my way through the introductory level biology labs—dissecting plant bodies, collecting caddisfly dwellings, identifying native Vermont birds and plants. I generally preferred the ecology-based classes, but I especially appreciated the iconic Drosophila Fly Lab, where we reared several generations of flies in order to monitor the reoccurrence of the "red," or wild-eye, type. When the time came to write up our lab reports, I always struggled with a tendency toward wordiness, forgetting the need to be precise within the rigid confines of scientific publication rules and standards. Abstract, materials and methods, results, discussion—there were no exceptions to the sacred scientific form.

For my final drosophila report, I experimented with the only wiggle room I had left—a catchy title—"Where Did You Get Those Wild Eyes?" Much to my dismay, I received a slash of red ink across my paper with a comment: "Science titles must be concise." As a young researcher, I felt my poetic sensibilities were drilled out of me until my ability to report was devoid of a unique voice, personality, or style.

One summer in college, I worked as a student research assistant in a study about bees and the phenomenon of "nectar robbing." The study actually *paid* me to watch bumblebees move from flower to flower—a job Emily Dickinson would have loved! My nursery training had prepared me for the monotony of the project: to study the circumstance in which, occasionally, a bee would avoid entering the mouth of a jewelweed flower, creeping, instead, behind the flower to nibble on the spur for nectar, thus avoiding its evolutionary responsibility to enter and leave the flower with pollen on its body. Our task was to figure out why the bees made this decision, and what its implications were for both host and visitor. I spent my time mesmerized, watching the smart bumblebees explore flower parts, nibbling their way into sweet nectar. I couldn't help but wonder how our presence, four researchers with tape recorders and gobs of coconut-scented suntan lotion, affected their movements. We went about our study as if we could pretend to be absent from the summer meadow, like shadows and phantoms, but I got the feeling those bees watched us as much as we watched them.

Although those long, hot days recording data reminded me of the pleasure of my time at the nursery, I felt I was at an uncomfortable distance from the natural world, even though I was standing in the middle of it.

⟜⊝

Another summer, I spent five weeks as a WWOOFer (Willing Worker on Organic Farm) in New Zealand. In exchange for four to six hours of labor per day, we were fed and offered housing for a short stay. Those were the general rules of the program, but they could be interpreted differently depending on what country you were in or whose land you were on. Often, we were more than cared for, treated as if we were long-lost family, almost. I was traveling with a boyfriend at the time, and we were able to hitchhike from farm to farm. Businessmen, clergy, truck drivers—New Zealanders are famous for their hospitality, and many of our volunteer chauffeurs would also invite us to stay over in their homes. A Maori man named Hehu took us out for fish and chips wrapped in newspaper before offering up his guest room.

During that trip I got a taste for living and working on small farms—first in Gisborne, then in Wellington, then in Nelson. There was the family with a large garden and a bay full of crayfish. We tended their garden all day and then helped prepare meals in exchange for a loft bed in an old carriage house and a seat at their long, wooden courtyard table. Every day they'd host a large group dinner with family and friends who lived either nearby or on some corner of their property. Dinners would often go on for hours, followed by music and sometimes dancing in the courtyard.

Next, we stayed on the south end of the north island with an older couple who raised pigs and horses (one of which served as the live model for the back end of the centaur on the TV show, *Xena: Warrior Princess*). Finally, there was the young couple with two small children. They grew tangelos and passionfruit and made baskets out of willow and lived on the wild edge of a national park. Watching them frolic across their fields, I couldn't help but feel like I had fallen into the life of the *National Geographic* photos I had coveted as a child. I wanted their life—I craved it—and when I returned home the idea of farming had firmly rooted itself in my body and mind.

Time is a funny thing. Years can be condensed into sentences or, sometimes, a day can feel like a lifetime. On 9/11/2001, my parents and I spent
the day circling through news stations. It seemed like there were new, un-
believable updates every few minutes and increasingly more disturbing
images of that gutted city. While we still didn't have a precise explanation
for why this was happening, it was clear that life as we knew it would never
be the same.

An obvious rift between my parents had been brewing since my years
in high school (maybe even earlier), but when I returned from college that
fall, I could sense the cool hardening between them. In my few weeks at
home after the shock of 9/11, the awkward silences were deafening. The
tensions were unavoidable and awkward and I was unsure of what to do. I
now knew why my sister had fled so decisively the minute her high school
diploma was in her hand. In the world, in New York, at home: everywhere
I looked, it was unclear what would happen next.

Home no longer felt like home, and all I knew was I needed to go.

Soil

THE NEIGHBORS' DOG IS YAPPING. Something has spooked the little Jack Russell, and now it can't stop itself. Their last Jack Russell got picked off by an eagle before I moved here. It was one of the first things they told me. Tenacious valley eagles breed, nest, and hunt throughout the flats. They'll eventually kill those neighbors' chickens too, one by one, leaving heaps of feathers and random bones in their backyard. During our first neighborly conversation, standing over our shared metal fence, they politely asked me to keep the grass down on my side. I smiled and nodded, suspecting it would always be unruly pasture.

The fertility of the soil in Skagit Valley is ranked within the top 2 percent in the world. Perfect for growing crops. However, the valley didn't become farmland until the late 1800s, when Dutch and Norwegian settlers hand-dug an extensive dike system that drained what was until then tidelands and marsh into arable soil. Farming here still thrives on the richness left from the past transient river channels and constant flooding. The visionary ecologist Aldo Leopold said, "Health is the capacity of the land for self-renewal. Conservation is our effort to understand and preserve this capacity." Now that river flow has been diverted and controlled, it's not yet clear what practices will become necessary to maintain soil fertility in the long term.

The number of farms and farmers in this valley has been dwindling due to operation costs and land prices, although, ironically, the farmed landscape is what continues to draw people. I confess, it's what made me want to stay: the beauty of a freshly planted field, the sunset slipping behind a green pasture of grazing cows, a serene gray heron posturing in a deep irrigation ditch.

To own a house on farmland in this county you need forty acres minimum, meaning you can't build a house unless there are at least forty acres of adjacent farmland. This code, established by a visionary group of farmers, helps to pace urbanization and was put in place to combat the ongoing development boom that started in the Seattle area thirty-some years ago. Some parcels, like my farm, were subdivided before the law was established and are only three or five acres—the perfect size for a country estate.

In order to have enough farmland, I rent some from a family potato farm to the south. My farm is bordered on three sides by potato fields, but very few other people in this part of the valley are trying to farm commercially. When I asked if it would ever be possible to buy a small piece of their land to increase my acreage they told me, definitively, that they never sell land, they only buy it. Renting is much cheaper than owning anyway, since land prices are so high in this area. It's a desirable place to be regardless of whether or not you farm.

My sheep barn borders the neighbors' tidy lawn on the north side of the property, and we have different ideas about how things should look around here. Those neighbors don't work in farming (former military and a nurse), although they do keep animals and appreciate the open space—as well as the associated tax breaks. They are nice neighbors, generally, but we don't have a lot in common. When the farmer across the street cut his corn (a rotation crop for potatoes), my neighbor lamented that her westward view was now too "dirty"—that is, the soil was exposed, and she found it unappealing.

The forty-acre minimum and other codes, as well as the county's conservation futures tax, which helps buy up development rights, preserves the land base in agriculture. Most long-term residents of this valley are on board with that idea, as indicated by the rash of clever bumper stickers created over the years: "It's Not Farmland Without Farming" or "Save Farmland, Live On A Hill." Landowners who have any excess acreage beyond their lawns usually just lease the rest to big potato, seed, or berry farms—or plant it for hay or feed, mainly corn or barley.

Land sharing is essential to all the large farm operations, as crop rotation has always been essential to agriculture. The soil is healthier for a myriad of reasons if you don't plant the same crop in the same spot year to year, and most farms need to have enough land to meet the market demand for their crops. Keeping rotations viable is essential to maintaining soil health and fending off a devastating crop disease that could result from continuous planting of one type or even related crops, like brassicas, in the same field year after year.

Definitions of soil health are relative; it depends on who you ask. It's one-third stewardship, one-third biology, one-third restraint. Agricultural soils have the ability to generate food and store carbon if they are managed right. However, the best way to manage soil is highly contentious here and around the world. Wendell Berry writes, "Because soil is alive, various and intricate, and because its processes yield more readily to imitation than to analysis, more readily to care than coercion, agriculture can never be an exact science." New ideas about soil health are slow to catch up with current practices because it takes years to improve your soil in any documentable, scientific sense.

The county has a "right to farm" act in place. Everyone who buys a new piece of land (farmers and non-farmers alike) has to sign it, agreeing they won't complain about the effects of agriculture in action. Annoyances like tractor noises or mud on the roads or prevalent animal smells. Although the number of active farms is shrinking in this county, agriculture is still the primary industry in this valley, and my farm is a part of that legacy.

I am not sure I have more of a right to be here than any of my non-farming neighbors, though. Eventually, I will get married and start a family, but for many years I will be a technically single, as noted on my land lease with the potato farm (Jessica Gigot, a single woman), experimenting with a wide variety of farm projects. Duck eggs, dried herbs, radicchio—ideas on how to make this farm financially viable. I imagine the neighbors will wonder what exactly is going on over here. I try to limit my judgement of them—like the lawn that looks like a putting green, the constant leaf-blowing and weed-whacking.

I get strange looks from the multigenerational farmers around us, too. They drive slowly and stare from their decked-out work trucks as I push the wheel hoe down each row of crops or fire up my small Kubota tractor, dune-buggy-like in comparison with their plus or minus 100-horsepower beasts. A retired farmer stopped by once and said, "You look like my grandpa when he was farming!" I laughed. I don't think it was a compliment; I don't think it was an insult either.

To farm at a small scale is to recalibrate the typical energy flow of a farm. Most of modern agriculture relies on faceless labor, paid (and treated) poorly, as well as high-end, never-fully-paid-off farming equipment. To farm small means, among other things, that the farm owner is one of the primary farm laborers. This used to be the way here in this valley, but now it's a novelty to see a farm owner bent over in a field. One advantage of small farming is that it gives a farmer the chance to know the soil well by tending it with regular care and attention.

Sometimes I brood over the fact that I am not "from" here, a history I share with many of my non-farming neighbors. I have had so much choice available to me in my life—maybe almost too much. I wonder. I do know that I decided this was the place where I wanted to be after years of searching and moving. On one of her visits to the Skagit Valley, before this farm was even an idea, my grandmother told me that this place was my "territory." She saw me here, and the life I might have, and through her eyes, I saw it too.

I am here because I continue to make the choice to be here, and, hopefully, I'm embarking on a way of life that will burrow beneath the surface and make roots of its own.

What does it really mean to own land? Yes, there are titles, but beyond the legalities of my tenure on this piece of ground, there is the moral responsibility to tend and conserve it, to make the place better than when I arrived, to steward the soil, making every effort to cultivate and preserve a fertileness that could foster food in the future. And I always remind myself that this farm wasn't always farmland, it was tideland. There is a longer history to reconcile with the present. One that I need to know and respect and consider in every decision I make. Coast Salish people have been here for thousands of years. In comparison, commercial farming is a fairly recent and possibly short-lived endeavor.

Soils have names. They can be classified into one of Earth's twelve soil orders, and then further subclassified all the way down to a specific, regional type, like Mount Vernon fine sandy loam, which is an example of one of the soil types that make up this farm. Many factors are used to describe a soil—for instance, its provenance, its texture, its chemical makeup. Soils develop over hundreds of thousands of years, a timeframe that our modern minds have trouble conceptualizing. In Washington, many of the soils are considered Alfisols, an order characterized by layers of volcanic ash built up over thousands of years. By contrast, the erosion and depletion of soil can happen within only a few farm generations.

Knowledge comes from experience, observation, and from staying in one place for a long time. Anthropologists studying an Indigenous community in Mexico found that the community had its own names for soil types, a unique classification system born solely from observation and daily use. The Indigenous system is astonishingly similar to Western scientific classifications.

⊹

A woman in her sixties pulls up to the farm stand that sits at the end of our driveway. I sell various produce and herbs at the farm stand, and I have just restocked it. It's late in the afternoon, and I'm organizing harvest crates outside when I hear her pull in.

"Hi, how can I help you?" (I don't normally come out to greet people, but I am nearby.)

The woman leans out her window and says, "You're living in my grandmother's house. Alice Nilsen."

I know that last name. There's a Nilsen Road pretty close by.

"I see you have sheep. My grandmother grew up on a cow dairy, but she raised sheep here too. Her husband was a dairyman. Her parents ran the dairy that was once here and they planted that horse chestnut," she says, pointing to the majestic tree on the other side of the driveway.

"Wow!" I exclaim, astonished and delighted.

"I'm excited to see what you're doing here and have been meaning to stop by."

We continue chatting. It seems no one in her family works in agriculture anymore. Several of her siblings have moved away to Idaho and California for work. She has one daughter in the area. The way she stares at the house seems filled with both nostalgia and relief. She seems grateful that someone has moved in and made it more farm-like, yet also disappointed it isn't her own daughter living here.

I am not sure what she imagines my life is like here now. I can hardly imagine what her grandmother's life was like in this farmhouse forty years ago. It's a good reminder that with each generation comes a new way of doing things. And change takes time. Maybe, somewhere, her grandmother and my grandmother are talking about me in this house, on this farm—and maybe they're both grinning.

Applegate Valley, 2002

MANY SMALL AND ORGANIC farms survive on seasonal interns. In some cases, the intern gets a stipend, free room and board, or some sort of exchange for organized learning and education on the farm. There are many creative scenarios that satisfy a farm's need for low labor costs and a newbie farmer's high learning curve. In most cases, the internship experience can be a positive first step into farming. However, many young farm interns end up putting in over ninety-hour weeks only to realize at the end of the season that they are lacking the essential resources, like money and confidence, to start a farm.

The vocation of farmer was not presented as an option in my young adult life. I was in my mid-twenties before I became aware of the land-grant universities in our country, agriculture education and research, and related government-funded cooperative extension services for farmers—the backbone of our rural communities.

While I didn't grow up on a farm, I did attend farm camp as a child. My second cousin, Margo, a Catholic nun, started a farm and creamery in upstate New York in the 1980s. As an eager first-grader on that farm, I made candles, spun wool, milked goats, and gallivanted with chickens. Most excitingly—to me—I got dirty. At the end of the day, I was covered in mud and manure. I remember distinctly that the feeling of joy and connectedness I felt on Margo's farm dissipated almost as soon as I passed out of the green field gate on my way back to my usual bubble of jazz dance classes, soccer practice, Happy Meals, and *Monkees* reruns on Nickelodeon.

Besides Margo, I never crossed paths with any full-time farmers. In college, I lived and studied in a bucolic setting in northern Vermont. Back in my home state, I didn't know any farmers either and never really thought of farming as a real profession, although Washington is a largely agricultural state. Now that I wanted to try farming, where would I start, beyond admiring a freshly plowed field or vibrant pumpkin patch from a car window? While I had some academic resources to turn to, I knew there was a lot to absorb alongside actual producers. After finishing college and leaving my parents' house, I decided to become a farm intern.

⊹

The village of Williams lies in a dead-end valley in southwestern Oregon, and when I arrived at the large two-story bungalow that was to be my temporary home for the summer, I noticed the mountains that surrounded us on all sides. I was twenty-two and starting an internship at a medicinal herb farm. I felt both safe and trapped by my new landscape. Williams is a few hours east of the town of Ashland, the last stop on I-5 before hitting the Siskiyou Mountains and descending into the vastness of California.

In the intern program, I agreed to work forty hours a week on the farm in exchange for housing, a food stipend, and a series of evening classes with local farmers, healers, and food producers. The other interns ranged in age from early twenties to forties; among us were a few aspiring farmers as well as herbalists. I was curious about how farms worked and what other lifestyles might provide alternatives to the way I was raised.

Alex owned the farm and herbal products business. He had started the company with his wife in the 1970s. They were divorced, but cordial, and they both lived on the property. I saw her occasionally sprinting to and from her outdoor shower in a tropical print robe, but otherwise she kept to herself. Alex, on the other hand, was very social when he was in town, and he invited the interns to use his saltwater swimming pool after work. Bathing suits were optional.

Coming from college and houses with roommates, I wasn't averse to living in a communal space. The interns' house had several rooms with bunk beds. I grabbed the room with one set of bunks and a view of the patch of forest that separated our house from Alex's mansion and pool. We would all travel a well-worn trail from our house to the farm that passed his oasis.

The downstairs of the intern house had an open floor plan that included a large living room and a long dining room table. The kitchen had every appliance one could imagine and a gray water system. There was a small office on the west end that held a library of books on herbalism and farming as well as a small computer. Outside was an outdoor shower and stone patio with a picnic table. We would often gather there at the end of the day and watch the light fade behind the horseshoe outline of mountains that surrounded us—the convergence of the Siskyou and the Coast Range.

Alex was a generous man and hosted several "kava" parties where he would tell us stories about his herb-scavenging adventures until we all fell into a dreamy poolside stupor. Kava, a plant in the pepper family, was one of the "super" products his company offered. The medicinally active compounds in kava, native to the South Pacific islands, are found in the plant's roots. Kava root, usually chewed, macerated into a juice, or drunk as a tea, was historically used in ceremonies for intoxication and to prevent anxiety. For me it did both and quite rapidly. As I drank the tea, my fingers and toes tingled as I felt the tension in my shoulders and back drift away into the night's still, warm air.

While Alex was off traveling to find new medicines or lecturing across the world, a warm-hearted band of dreadlocked white guys ran the farm. They had all migrated here to live off the grid, and they all loved every aspect of farm work. I suspect they also loved the freedom Alex gave them to manage the fields. The farm essentially broke even financially (the tinctures were Alex's bread and butter), but the bare-bones crew took pride in every step of growing, from cultivating the tiny greenhouse seedlings to drying the harvested herb plants. Over a hundred different types of plants were grown on the farm, and there were several acres of display gardens.

The days in Williams were hot, dry, and long. Temperatures could get into the hundreds, but wonderful shade trees encircled the farm, and a creek ran along the west side. The workday was from eight to three, and then we would have a few hours to swim and cool off before we had an evening class with a local expert. That was the "study" part of the "work-study" trade we had all signed up for, and our evening sessions brought in a whole host of local characters, from a charismatic herbalist/seed grower to an intensely focused vegetable farmer named Sensei. There was also a local naturopathic doctor who gave a series of herb-related courses, and she offered us work-trade at her horse farm in exchange for a visit to her clinic in Grants Pass.

Chris, a fiery but gentle-hearted field manager, told us this area was considered sacred by regional Native American tribes like the Takelma, and occasionally one of the interns would find an arrowhead in the production fields. Chris had accumulated a collection of them on the dusty dashboard of the pickup truck he used to drive the interns around the farm. He shared bits and pieces of local botany with us as we worked. As an aspiring herbalist himself, he always handled the plants as if they were his own precious children.

Chris was also on gopher patrol. Gophers were one of the major pests on the farm. One afternoon an intern found a cache of mixed medicinal root pieces, a private vermin dispensary, stashed away in the far corner of the field. The thrill of discovery, the questions of what roots they chose and why, sent Chris into a tizzy. We couldn't distinguish what each plant was, but there must have been at least ten different types of root crops represented, such as dandelion, red root, and burdock.

Chris rubbed the top of his head and leaned up against his truck.

"Damn it!" He realized what this meant—the gophers were at work in many parts of the farm making their hidden medicine middens, damaging all the precious medicinal crops. He dashed off in a fury.

Echoes of dynamite resonated throughout the farm all that day as he tried to send the tenacious critters an emphatic eviction notice.

The challenge of farm work was humbling. I had never done field-scale cultivation, hand harvesting down hundred-foot rows of calendula, hoeing for hours through a half-acre planting of rue. I was surrounded by a new host of crops, far more diverse than peas and carrots. It took quite a lot of discipline to weed around the bright-faced dandelions, but it turned out that dandelion root was prized as medicine and one of the more profitable crops for the company.

The farm's most important crop financially was *Echinacea*. Taken as a tincture or tea, the plant was becoming one of the Western world's most important cold and flu remedies and immune system boosters. While all parts of the plant are used, the root is the most prized for its healing compounds. After the flowers have bloomed, the plant begins to store all its energy and resources, the precious secondary compounds that we use for medicine, in the roots. As Chris drove the tractor with a modified potato digger on the back, muddy root balls flew through the air. The interns were in charge of catching or chasing each one, shaking off the dirt clods and piling them in black plastic crates for washing. After I got used to the confusion of the flailing plant parts and soil clod shower, this job became quite hysterical, and our crew of interns ran around like children, laughing and covered in dirt.

Some days we were broken up into crews. I was picked to hand harvest oats with some of the dreadlocked crew. Jim, an older hippie, was our crew leader. He was smaller than Chris and loved to talk. I appreciated his commentary and the way, when he had an important point to make, he would stand up in the field, shake his long mane of dreadlocks, and adjust his round-framed glasses until they sat right on the bridge of his nose.

A cool fog hung over the valley, and we were talking about the work of farming. Jim had lived a previous life as an activist (and most likely an anarchist) before becoming a full-time farmer. We were all hunched over the oats, our open palms sieving the immature oat seed heads from the tall golden stalks. Jim stood and arched his back.

"You see, man. White people can do this work. Why more white people don't farm is beyond me. Even these big organic farms, the Hispanics do all the actual farm work, but they put pictures of white people on the labels and ads. That is why I like working here."

He adjusted his glasses again and at that moment stared at me directly.

"It's not fair what we pay all these Mexicans. They grow all our food because we are too lazy to do the work and we can't even bother to pay them well."

⊷

I had been to Mexico, once, for an internship in college. A friend and fellow sophomore and I spent the month of January in Colola, a small coastal village, helping with a sea turtle protection program. It was idyllic, and I fell in love with all the families and friends I met there; the hospitality was palpable.

Our group of volunteers ate with a family in the village. They had a large cistern for washing dishes outside and a small kitchen inside their home next to a room full of beds. The bathroom was down the road. Cooking was good income for them, but they offered so much more than food. The family—three daughters, their mother and father, and a grandmother and aunt—welcomed us in with open arms at every meal. They asked about our lives and our families. They told the best jokes that I could almost understand even with my basic Spanish.

The food was served on the long table outside under a makeshift canopy. It was simple—chicken, rice, sometimes beans or avocados from the tree next door, and tortillas, lots of warm, handmade tortillas. The grandmother and mother would wake at dawn to form the dough, and the daughters would help press it. Corn, water, salt, love—it was the best food I had ever tasted.

After five weeks of following baby turtles, harvesting coconuts, and savoring the most delicious food I'd ever eaten, it was time to go home. Teary-eyed, I promised to return and visit very soon. Several of the men laughed as they told me they were headed to Wenatchee (a town a few hours away from my hometown) in a few months to work. Almost half of the village earned their yearly income in Washington apple orchards, making the harrowing trek over the border in springtime and back home for Christmas each year.

I had never thought about who was growing my food.

I would pack up my dusty, small backpack and return to college life, the convenience of cafeteria food, late-night parties, and frigid morning walks to class. Snow would surely be piling up around campus when I returned, but the warmth of this beach, this place, would stay with me.

❧

As the summer progressed at the herb farm, I realized I did love the physical work, especially the still moments of concentration when we all ran out of things to chat about. I was good at it. I appreciated each ounce of water

I glugged during our breaks and the occasional cool breeze that passed through the valley and over my sweaty neck. Even though I was athletic growing up, my body had never felt so fully engaged. The sun glared down, but I found a sense of calm and centeredness I had never known before.

Before that experience I was illiterate about farming. I hadn't known how much work it took to grow food or medicine. I took pride in my daily tasks as an intern. I was also astonished by my ability to sleep deeply, my body sore and aching most nights, in foreign places. Before nine I was in my bed, falling asleep with a noticeable tight and tense sensation in my back and hamstrings from a day of manual labor. I didn't miss going out at night, and I rarely went to town. All I could do was lay my body down.

The physical exertion was healthy. At the time it felt like I was made for this work. Laboring this way, after years of studying, brought me into the present moment and I experienced a deep connection to my body and the Earth. I noticed the position of the sun in the sky, its trajectory. I saw different leaf shapes and colors. Childlike wonder, rushing in like spring river water, overwhelmed my senses. While learning about the plant medicines was fascinating, the visceral sensations of being around alive and verdant plants all day was a whole other medicine entirely that I hadn't known I needed.

❧

Living in a house with eleven people, all curious about and comfortable with the philosophies of natural medicine and sustainable farming, was a new experience for me. My roommate Mel and I bonded immediately, since we both stemmed from Catholic, midwestern roots, and we shared a certain cynicism about the eccentric beliefs and Earth-worshipping culture of Williams. We both had been biology majors in college, and I think that's what helped us stay grounded amid all of the new and esoteric learning we were experiencing. We were also well-matched pace-wise in the field. Mel had interned at a vegetable farm in Colorado before arriving at the herb farm, and for my part, I was surprised by how fast I could work—and how often the crew would pick me to help them on special projects.

The rest of our crew had arrived from all over the country. Most of the other interns had no farming experience. Two girls from the South were recent high school graduates with an interest in herbs and yoga. In our house we had a raw foodist from New York with a desire to be "out west"; a transient herbalist named Germaine who loved to make life decisions by putting things "out to the universe"; another east-coaster with a propensity for taking acid or performing homemade colonics in our shared bathroom; and an Oregon native who wanted to learn about herbs for her baby. (She was not pregnant yet, but was excited about the prospect of becoming a mother at some point in the future.)

For the most part we all got along, but as the heat of summer weighed on people's patience, our sense of courtesy faded. We became comfortable with each other, and our politeness soon resembled sibling banter. We had a stipend for the house to buy groceries, and we all agreed to cook evening meals together, each intern taking the lead in cooking in turns. I tried my first raw food pudding and many interesting nutritional yeast dishes that summer. Sometimes the vegan or raw dinner options failed to satisfy our overworked bodies, and Mel and I would secretly take the house bikes to the general store a few miles away and buy chocolate ice cream.

I turned twenty-three at this herb farm. On my birthday we coincidentally had an "educational sweat lodge" by the creek behind the fields. This was one of our many classes, which ranged from seaweed harvesting on the coast to seed saving with an eccentric seed smith named Luca. With his long, gray hair, he resembled an aging rock star, and in actuality he really was a star of sorts—in herbal seed circles. People knew about him around the world, and he traveled in the off-season to teach and sell his seeds. Years later, I continue to buy seeds from him for our farm.

The structure for the sweat lodge was a simple dome made out of riparian twigs and branches covered in canvas. Kathleen, our neighboring naturopathic doctor and teacher, led the ceremony and the preceding class about detoxification rituals. One of the farm crew tended the firestones used to heat the lodge. They'd both been at the creek all day collecting suitable rocks, and I arrived early and watched as they loaded the hottest rocks into our fabric-covered stick dome. As Kathleen poured cool stream water onto each stone, the steam filled the space with a pungent, wet heat.

Sweat poured down my face after only a few minutes in the dome. I felt connected to another world I didn't know existed. The simplicity of rock and heat, fire and water, was powerful, even in our non-Indigenous, Western-hippie sweat lodge. I could understand how so many cultures found sweat lodges powerful, even religious. The southern girls had to step out early, overwhelmed by the spectacle of it. And the oppressive heat. Germaine, having experienced this before in her nomadic wanderings, was already in joyous sobs and tears with her arms raised and open straight up to the top of the dome, as if she was touching the sky. I never officially blacked out that evening, but as Kathleen invited us to go deeper into ourselves, I could hear Mel snickering behind me, and then nothing. There was a calm silence and then a deep, unnerving opening in time and space.

I couldn't tell if I was crying or not because of all the sweat on my face. As I continued to look into the void, I saw myself in the fields, bent over weeding; I saw the delicate *Echinacea* flowers swaying in the wind. Hands moving faster and faster. The planet spinning faster and faster. I was dizzy-drunk and soon on my back against the sandy riverbank.

When I opened my eyes, Kathleen was fanning the doorway and reminding us to drink water. I had never seen our crew so sweaty or dusty, even after the longest day in the field. When I stepped out of the sweltering shelter and into the creek, I heard the rustling of willow limbs and watched the moon reflect off the corn leaves (the farm harvested corn silk for medicine) in the back field. It was just after sunset, and the coolness of the water on my feet returned me to myself. Or the new me I was becoming.

⊕

On the herb farm, while each of us considered interning to be part of a sort of life experiment, we took ourselves very seriously. I noticed how competitive I had become with my pace of work. The farm work enlivened my body, but it often took brute force and struggle to get through a day on the farm. As the weeks went on, our group vacillated in its vigor. The southern girls were often sick and needed to take a day off. Our raw foodist seemed to slow down as the heat rose, and we'd find him slumped in the shade of a Garry oak, gorging on his soaked nut supply frosted with blue-green algae. One quiet man in his mid-twenties disappeared altogether.

Our remaining weeks on the farm were bittersweet. Some people were exhausted and more than ready to return to their old lives. As we entered into our final days, we celebrated with a last house meal of tofu stir-fry and avocado chocolate pudding. We all promised to stay in touch, but in truth, I never saw any of these people again. Just like I never made the avocado pudding, despite having asked for the recipe.

Some of the farm employees had hinted to me about the need for an assistant manager on the farm's display garden, but the thought of being in Williams indefinitely was confusing. I felt a new sense of my body and its capacity. I loved the physical work of farming and the rhythms of working outside, but I wasn't ready to be anywhere. I had followed my camp counselor love-interest to this part of Oregon; we had been seeing each other on some weekends during the internship and were talking about what was next for us.

I was still feeling unsure and unsettled. I appreciated the farm crew and the raw beauty of this valley, but the intense, homestead ethic of Williams' permanent residents was alien to me. The idea of going out to a movie and using store-bought soap sounded good. By the end of the summer, I headed out of the curving Applegate Valley roads back to civilization. I felt physically vibrant, but still terribly lost.

Animal

I LIFT THE LID OFF the silver trash can, where the grain is kept safe from rats and racoons. The ducks know this sound. There is movement in the small, messily painted house followed by laugh-like cackles. The ducks are hungry. They march down the ramp and form a circle around the feed trough, flapping their webbed feet into the fresh mud. The click of the gate ignites a crescendo of want. The ducks are really hungry now, and assertive. I try to spread the grain evenly as they dive in, picking up every morsel with their smooth, ellipsoid beaks.

Sometimes, when I am feeling generous, I gather kale leaves and salad greens from the garden beds, the ones the slugs got to first, and toss them over the side of the run. The ducks scramble and fight over the fresh stuff, playing tug-of-war with the thick, green shards.

I ordered the ducks online. There were so many kinds to choose from I had a hard time deciding. White and fluffy and regal; tall with brown wings; or scruffy brown (reliable) egg producers. I was dreaming of their big, rich eggs. I limited myself to a dozen Khaki Campbells (the reliable ones) because I didn't know where they would go yet or how many would be males. I hoped for only females and maybe one male since fertilized eggs were supposed to be healthier. However, you only want one male for every ten females or so. The males get rough if there is too much competition.

Driving to the post office to pick them up, my heart pounded. What if they were all dead? That would be a disaster. Our post office is out in the middle of nowhere, in front of the gun shop. The staff are not fazed by odd packages filled with chicks, or wool, or whatever gets shipped out of (or into) this valley. They have seen it all, I imagine. I grab my quacking box and go.

At home I have prepared a kiddie swimming pool and a heat lamp for them in the barn. I've covered the bottom of the pool with pine shavings. One by one I lower them down with a cupped hand. They run in wild circles before hovering together in the corner. They all look healthy and soft. I can't sex them yet; I'll have to wait a few more weeks to sort the males from the females, watching carefully to see where a curled tail feather forms—the sure sign of a drake.

⚬

I bought the first ten sheep before even being on my farm. A famous cheesemaker in eastern Washington, faced with her husband's failing health, a food safety violation, and a looming, long winter, needed to retire and rehome her sheep quickly. Lorraine, a farmer on Whidbey Island, an hour south of here, and I decided to buy the sheep together. She agreed to pick them up and bring them back to the west side. She would keep most and when I was ready—housing, bedding, water trough—I could pick up my ten. It was beyond generous, especially since at the time I made the deal I had no idea how to even haul livestock.

To get to Whidbey Island you can take a ferry or, if you are coming from the north, you can drive over one of Washington's tallest bridges at Deception Pass. The one that looms over 185 feet high and sometimes shifts slightly in the wind. Where people occasionally decide to cascade to a hard ocean death in the sound. I can never quite look all the way down when we pass over it.

I employed a friend to follow me with his trailer. I led him to the center of the islands and then down a long chain of dirt roads. He was nervous. I could tell from the rearview mirror. I saw him looking left and right, hardly believing that down these roads we'd arrive at sheep. A psycho-killer, maybe, but not a farm. Eventually, we came to a clearing, a majestic barn, and a sloping view of the ocean. The sheep were penned up in the lower barn, and we loaded them one by one. Bertha, the largest and the last to load, was pregnant.

When we arrived at my farm, the sheep scuttled to a corner of the paddock. My friend had brought his two young boys, and they danced around the trailer, proud of their dad for helping me and enamored with the new animals. The sheep were all sizes and colors, a motley crew of brown and white and speckled faces. We knew they were a milk breed, but we had no idea what we'd do with them besides use them to keep the grass down.

When I told my neighbor down the road, a six-foot Texan who raises dairy goats, that I got some sheep, she asked, "How many?"

"Ten," I said proudly.

"Ten!" she yelled back at me. "Well, you got your hands full now!"

↢

I only had house pets growing up, so I joke that the sheep are my adult 4-H project. Once I am in their presence, I cannot help but admire their power and strength. After a lifetime of seeing sheep only from the roadside or in pictures, I have them now up close and in the pasture. But the joy I feel at the sight and sound of their romps and baa's is coupled with an unbearable weight. I am now responsible for these creatures.

On that first day, it quickly became clear that Bertha, with her dark face, large oval eyes, and cream-colored fleece, was in charge of the crew. She bellowed and they all ran to the right; she barged through and they all made way for her. In no time at all, the new social order was established. I watched them explore every nook of their new space.

↢

Initially on the farm, I had only my small Subaru hatchback. It was a great car for adventuring in the mountains, but it proved to be less useful on the farm, despite its all-wheel drive. Before the sheep arrived, I scrambled to get all the supplies needed to care for them. I crammed three bales of straw for their bedding into the back of that car. The smell was so overpowering I had to roll down my window. As I drove north on I-5, little yellow confetti blew all over the place and out the window. I had to move my seat all the way forward to make room for the long eighty-pound bales. Hunched over the steering wheel, I probably looked like I was driving a backwoods clown car.

At a stoplight, I realized a friend from town, a wildlife biologist, was staring at me. Initially he didn't realize it was me amid the golden mess. When he recognized my face and hair covered in straw, he burst out laughing.

"Hi, Jess!" he yelled, still laughing.

I chuckled, "Hi! It's for the sheep!" We waved and I sped off at the green light.

It took me weeks to clean all the little bits out of the car's nooks and crannies. I knew that the next step, after securing a piece of farmland, was to get a farm truck. Something reliable, maybe older than me, something that could take a beating, haul heavy things, and wouldn't need to be kept clean.

One day I was driving to a nursery to get some shrubs to make a new hedgerow—a wall of plants that could act as habitat for native flora and also protect our plants and animals from the drift of pesticides from neighboring fields and gusts of damaging winds. An old white Ford F-150 with a mauve canopy drove past me in the opposite direction, and I remember thinking that was exactly what I needed. After the nursery, I headed back home and saw that very same truck parked on the corner with a For Sale sign in the window. I called the owner immediately and knew it was fate—he lived just down the road and was home! I named her Tina because of her big wheels. The front windshield is cracked and the seat upholstery is torn, but she does her job.

Over the years Dean—my boyfriend when I moved onto the farm—and I have used the truck to haul animals, get feed, and make regular trips to the dump with old boxes and bags and scrap plastic from the fields and high tunnel. Tina is not easy to clean out. It's a pain to lift the heavy canopy on and off, so we normally leave it on. You have to hop up over the tailgate and, on your knees, sweep it all out. Feathers, hay, old grain bits, strings from feedbag ends. Everything ends up there.

⊷

When I bought a new ram from a farm outside Chehalis, he rode home in the back of the farm truck. He was quiet and relaxed, even when I took him through a Burgerville drive-through, though he refused the fries I bought for him. Most animals are too nervous to eat in transition anyway. I, however, was not. I inhaled my burger, fries, and a chocolate hazelnut shake, running countless "what could go wrong with a woman and a ram in a truck" scenarios through my head as I drove.

After acquiring several animals during the first year on the farm, it occurred to me that I might never leave again. Not with all the mouths to feed. I didn't have kids yet, nor any real inclinations toward mothering, and up until that point, I had cared for only one dog on my own in my young adult life. Surprisingly, it was easy to find farm sitters, folks who wanted a weekend getaway in a quiet, bucolic setting.

One woman in particular was quite enamored with the farm animals, including the dog and cat. She worked at a nearby bakery and often agreed to stay and farm-sit, cleaning the entire house from top to bottom during her stay. One day she stopped by after work to tell me her aunt was looking to rehome a few animals. A few, it turned out, meant a lot. Her aunt was single, and she loved animals and had access to land. Based on my own accumulating assortment of creatures, I'd begun to understand how a menagerie could manifest itself in little time. Sadly, the truth was the aunt was sick, and her family was trying to rehome her animals quickly.

Standing in our kitchen, the woman asked, "Do you want any horses?"

I shrugged, considering the possibility, having been on a horse only a handful of times in my life.

"They're too big and I am not sure what I would do with them. I'd have to build a new barn."

"What about mini donkeys?"

My heart leapt. There was a mini donkey farm on the other side of the valley, and every time I drove by, I fell in love with their gray, scruffy bodies. I had read that donkeys were common guard animals for sheep. Many sheep farms had either donkeys or llamas.

In high school, my best friend's mom had a llama. That creature was unpredictably moody and could spit, so I always kept my distance. My friend's mom rode horses, and her family always kept a motley crew of equines and dogs and cats around their place (as well as the llama). I spent a lot of time there, taking in all the creatures and chaos. Their farm, tucked into the forest with a barn and a horse ring and a large pond my friend's dad had dug himself, was always abuzz with life and action. By comparison, my family's house was pretty austere, except for the warm presence of one mild-mannered spaniel.

My response was immediate. "Yes! I want the donkeys!"

A few weeks later the pair of donkeys arrived at the farm from south of Seattle. The aunt was already on hospice, more ill than I had realized, and her sister delivered them to us. The donkeys, stoic at first, were hesitant when we slowly opened the back gate. She had clipped a thick rope to their halters and used it to guide them to the back of the property. They were reluctant and dug their hooves into the gravel. Their smooth, black-lined backs only came up to my hipline, but compared to the sheep they seemed huge and unmanageable. If they had reared or bolted, I would have no idea what to do.

Audrey and July are sister and brother, and they were twelve and ten years old when they arrived here. I would learn later that donkeys live to be over forty years old. When we released them into their field, unlike the sheep, they didn't stray far. In fact, they stayed put and stared at us with long, knowing faces. After a few minutes of stillness, both donkeys moved toward me. I held some carrot pieces in my pocket and they knew it. Audrey nuzzled my wrist. Her stiff coat tickled my skin. I offered them my gift, a rare treat; donkeys really need only hay and water. Each gobbled up the pieces with smooth, thick donkey lips. We bonded for life in that moment. Or, more accurately, I became hopelessly smitten.

I didn't expect to become an animal hoarder, but it is amazing what a little open space will do to a person. I had always been a plant person, had studied only plants in school, and I'd been intending to grow only plants, vegetables, and herbs on the farm. In many ways the animals were an afterthought, but once the sheep, donkeys, and ducks got settled, I couldn't imagine the farm without them. The farm animals started to feel like my first children, and also my teachers. They would appear in my dreams, and I found myself wanting to be outside and around them as much as I could.

Despite my exhaustive efforts to care for the animals and understand their needs like a naïve new mother, strange health issues have come up. Things I would never have expected to deal with—like the paralyzed duck who needed to be tube-fed charcoal to clean out its stomach of toxins. It lived, after an emergency trip to the expensive after-hours vet. I remember driving with the duck by my side; it was wrapped in a towel and tucked in an old file box. My heart beat hard the whole way to the vet. I didn't want the duck to die because of my ignorance, and I knew there must be an explanation. However, that emergency was a good reminder of how much I didn't know, how quickly things can go wrong.

Then there was the poor lamb that got its ears stuck in the wire fencing that outlines the paddocks and needed to have them amputated. She had a black coat, so I didn't notice the blood when she injured herself. At some point, when dangling her head out of one of the little squares of fence to reach some fresh grass, she'd managed to cut the underside of both of her soft oblong ears on the wire edges, where the pieces intertwined. I'd been wondering why they were suddenly hanging so low. I learned very quickly to never let a question like that linger—when something was off with an animal it meant something serious.

When I grabbed her and held her panting body between my arms, I bent over and lifted one ear. Maggots! They were crawling everywhere: plump, pulsing white bodies were burrowing in and out of her flesh. I screamed, let her go, and jumped up, and she ran to the far end of the paddock. I was sure I would have to put her down.

After the vet inspected the lamb, he gave me two choices: euthanize or amputate. Thankfully, amputating the ears was cheaper. She was a survivor and she knew it; Grace grew to become one of my most beloved animals.

My urge to acquire so many animals surprised me. I grew up with Sundae and maybe a goldfish, and two gerbils who died because they ate their wheel. The farm soon became animated not just by freshly tilled earth and proliferating plant life, but now these four-legged creatures too. Creatures that watched me as much as I watched them. Their sounds and even smells were nourishing.

Bertha, who arrived here pregnant, was the first sheep to lamb on this land. Two babies popped out early on a May morning while I was sleeping. They were still slimy and stumbling when I made my way out to the barn. Bertha doted on them, an experienced mother, and they lived. At that point I was grossly unprepared to help her had something gone wrong, like a breech birth. I look back at my cavalier attitude now with a mixture of shame and humor. I take all my animal care much more seriously now, knowing so much more about what could, in a heartbeat, go wrong.

At the end of her long life, Bertha died of natural causes on the farm. I found her motionless on Ash Wednesday—she'd died in her sleep. I cried that day and many days after. I had lost a teacher and friend—the wooly mirror that showed me how much I was changing.

⌖

There is one feed mill in this valley. The big farms keep it in business, but a smattering of small farms and horse farms get all their forage there as well. As a small farm, we are grateful for all the "big guy" business here, like the grain mill and the tractor store, because it makes it easier for us to get the supplies we need. We know we are not the ones keeping these businesses running, but I'd like to believe we'll be considered valued customers over time. Or, if there are enough small farms, perhaps we could, combined, equal the buying power of one potato family.

I get off on the last exit in south Skagit County and head west through the roundabout. Parking outside of the office door, sandwiched between the processing building made of cement and the rectangular storage shed, I run in to make my order and pay. As I wait by the car, I stare up at the huge galvanized steel silos looming over me. This is where the grains, like corn and soybeans (from elsewhere) and wheat and barley (from here) are stored.

A shy boy, barely in high school, drives the forklift toward the back of the truck. I open the tailgate and thank him, and he rapidly stacks fifty-pound bags in straight piles. He nods and raises his almost-man hand that speaks for him: "You're welcome, and goodbye ma'am."

Lopez Island, 2003

MANY PEOPLE CALLED HIM a zealot, but I initially found Gerry to be like an old, spirited grandfather. My second farm internship was on his small homestead farm in the San Juan Islands. I turned twenty-four the month I arrived on the islands, just as all the leaves and blossoms were opening. After the herb farm, I had taken a brief office job in southern Oregon. However, after taking a Master Gardener course over the winter at the local extension office in Central Point, I was itching to be on a farm again. I left that chapter, and a budding relationship, behind.

Gerry had a lot of ideas and plans for his farm, and his age, somewhere beyond retirement, was irrelevant in light of his dreams. While still teaching at a college, he had built his own barn on Lopez Island and paid off the main costs of his land. When he became a full-time resident there, he cultivated his own CSA (community-supported agriculture) with families from the island who came by weekly for produce, milk, and meat. They would pay him a lump sum at the beginning of the season and in return would receive a weekly box of fresh food items. He loved having an audience on the farm, and he welcomed each CSA member in as if they were one of his university students. Some people are natural teachers.

Gerry and I were similar in height, about five feet seven, but his body had a stocky box shape, evidence of his strong German ancestry. His wife,

Marie, was younger and tall with soft, brown hair, and both of them had gentle faces and rough, thick hands that revealed the hours of work they put in daily on the farm. Marie had also been in academia and was now the island's English teacher. While they sported Carhartt pants and thick farm coats, there was still a properness to their demeanor. With a little wardrobe change and sprucing, it wasn't hard to imagine them at the front of a large lecture hall.

✢

Lopez is an idyllic island—rolling, green hills; pristine coves outlined by rocky cliffs; blue water lapping along the shore—and I was ready to dive in. As we worked together, Gerry could never turn his inner teacher off, explaining how and why he had set things up the way he had and exactly how he was making money. I appreciated the breadth of knowledge he'd developed. The internship with Gerry had a very regimented learning component to it, and he and I developed a reading list for the summer as well as a research project for me. The creative and eclectic island community had an extensive library of books on farming methods and philosophy that I devoured throughout that summer.

Gerry and his wife Marie took the ritual of their work seriously, and their farm was in many ways a pastoral dream from another era. The summer I was there I joined their annual celebration of the summer solstice. We toasted the longest day with champagne and a strawberry, a Norwegian tradition Gerry had acquired in his academic travels. I listened contentedly to them recite Dylan Thomas's "Fern Hill" in its entirety.

✢

Our routine was defined early on, and we rarely strayed from it. On the farm we observed the Sabbath, but every other day was a workday. The farm was nestled into the central part of the island and surrounded by trees with a sliver of a view of the Puget Sound. Their life was sublime in many ways, the actualization of a vision they'd had in their previous urban life in Seattle.

Gerry and I spent countless hours weeding in the gardens, cleaning the chicken coop, and milking their cow. His silvery-white hair, showing no signs of thinning, would toss in the wind. Compared to the herb farm, which included days of working on the same crop or activity, my daily chores at the homestead were diverse. There was a multitude of things to do and creatures who required daily attention.

Animal husbandry was new to me; I became quite fond of Loveday, the cow who provided milk, cheese, and yogurt for us and several neighbors. It took me a while to get past my feeling of intimidation around her general girth (and the bang of her leg hitting our single metal milk bucket). However, after a few weeks I got into a nice rhythm of milking, cleaning stalls, feeding animals, planting, harvesting, and walking the land.

A previous intern had helped Gerry build the straw-bale house. It was cool to the touch even on the hottest days. Stepping inside, I felt like I was entering a womb of sorts. My bed was in the loft, and downstairs there was a couch and small table and desk. It was a welcoming space, and I enjoyed looking at the farm from the small, rounded windows on either side. Often, I was compelled to put both hands on the wall just to feel the earth. It was similar to impulses at the beach, when I wanted to let my bare feet sink into wet sand. There was safety in feeling connected to all of the layers of earth below. On Lopez there was a strong sense of a relaxed island life—people hitchhiked and no one really locked their doors—and I reveled in the smell of new terrain.

❧

As far as the toilet was concerned, there was one in the garage—along with a little sink and shower. Excessive water usage was discouraged. However, Gerry took a lot of pride in the "humanure" system he had created in the woods near the straw-bale house. I was expected to use it for most of my business.

"You want me to use that bucket?" I asked in disbelief.

He nodded with pride. He had built this three-chambered composting system with the help of a previous intern.

Gerry gracefully elaborated on the global benefits of humanure and the ancient Chinese fertilization practices involving the nutrient-rich compost as he passed off the clean white plastic bucket. He could be very convincing.

I lay in bed one night and read the literature he had given me about humanure. *I guess I should give it a try. I guess I can poop in a bucket all summer.* Layers of sawdust were presumed to cut the smell, and he had a three-bay system for rotating the processed material from year to year. You dumped the bucket of wood chips and shit into the first pile, which was turned regularly with a pitchfork. Once that material had broken down, it was moved into the next bay, and so on. The third pile, the aged humanure, was monitored for its temperature. If it hit the target temperatures, it was deemed ready for use. Age and time determine compost readiness. Heat can kill pathogens. After a week of solid humanuring there wasn't a pungent odor, but it was still a hard process to get used to, especially the weekly bucket clean-out. I had to deal with my own shit in a way I'd never had to before.

On that quiet island, I savored the warm nights and spent them reading and writing. It was the first time in my life I had lived alone. While I was there, I received a small stipend from a grant the farm had been awarded through the state's land-grant college, Washington State University. This was considered a hybrid internship; it paid the farm and me and offered training and lodging. I had free room and board and all meals provided. I was able to take Sunday afternoons off, which was my only time to have contact with people on the island or back home.

After a summer surrounded by people in a shared house at the herb farm, I both rejoiced and sulked in my solitude. A few weeks after I arrived, the island's community center hosted the screening of a documentary called *Hot Potatoes*. The film was about the potato disease, late blight, that had caused the Irish Potato Famine, and how that same disease is managed today. I hadn't met many people except for the CSA customers or the occasional neighbor who would stop by the farm to talk with Gerry. I knew I wanted to go.

As I entered the community center, I immediately felt welcomed. The hosts, a group of volunteers, were setting up chairs and a table of tea, water, and cookies in the back of the room. I lingered by the table, watching all the people arriving. Everyone seemed to know each other. As I took a bite of cookie, a few animated locals realized I was a new face and came over to greet me and ask me about my time at Gerry's farm. The conversation was refreshing, and I almost didn't want to stop to watch the movie. I settled into my seat and was soon whisked away to the highlands of Peru, the potato's place of origin.

The movie was surprisingly interesting. What was plant pathology? There were diseases of plants? The film documented the struggles of current US potato farms, as well as subsistence Andean farms, in their efforts to keep their tuber crops healthy. Many of the same serious plant diseases, such as late blight, are shared across the globe. The documentary outlined the abuse of chemical controls as well as new efforts to breed a potato resistant to late blight, a virulent disease of potatoes that can kill a plant in a matter of days. One of the interviewees in the film was a Washington State University scientist based in Mount Vernon, the town you pass through to get to the San Juan Island ferries. I had no idea there was so much agriculture in that valley—or that potatoes could get sick.

When I returned to the farm, I pondered the idea of an agricultural research center. What else did they do there? Intrigued, I made a note to look it up online next time I was at the island's library. In the meantime, I spent a good half hour looking at the stars by the pond, thinking about the new faces I met, before retreating to the straw-bale house. The time away had been much needed. Now the reflection of the stars in the calm water soothed me. I could hear the far-off splashing of our resident river otter and fell asleep to the calming song of a loon. After that dose of social interaction, I was glad to be back by myself and grateful to be on the island.

⟡

The bean lady showed up a few weeks later. She was a professor at Washington State University in the vegetable horticulture program. She had collaborated with Gerry in the past and was on the farm to test out some bean seeds for a disease called halo blight, cause by a fungus that moves from seed to leaves, leaving a large round ring on the leaf that resembles a halo. This affects the overall vigor of the plant and can reduce yields. At that point I was intrigued by the idea of research in agriculture. My experience in college had been with native forests, rivers, and meadows, studying wild and indigenous populations of bumblebees. Could you really do science in a man-made landscape like a farm field?

She had short brown hair and a straightforward personality. I could see there was potential for her and Gerry to butt heads. They challenged and intimidated each other. She showed up with a clipboard and her ruler and got right to work measuring the height of the beans Gerry had planted earlier that spring. He made a few comments about where she parked and that she was a little late, but she did not flinch in her focus on the beans. He went back to the house while we worked in the field.

Gerry had mentioned to her that I had a background in biology, and she enlisted me for the summer to measure plant height and observe any unusual symptoms in the plants. That type of research was interesting to me, and I was happy to have been volunteered for bean duty. I had been working on a compost project for the farm, documenting the temperature and content of the piles, but now I was collecting data. She was relieved to find someone with some science experience on the farm and we worked well together.

⊷

That summer I didn't have many other visitors. My mom came by for an afternoon with some of her friends. My dad and sister visited me on one of my Sundays off. They were supportive, but also a bit unsure of what I was doing with my life at that point. Later in the summer a friend from college, Kate, came up for a weekend, and I was given Saturday night and all of Sunday off for her visit. It was a special treat to have a longer amount of time off to spend with a familiar face my age. Gerry was usually strict about my hours, but I think he could detect my growing loneliness and felt sorry for me.

My friend had biked from her home in the north end of Seattle to the train station and from the train to the ferry terminal, a substantial ride. I had imagined we would spend the day biking the island, something I had wanted to do all summer. Unfortunately, when she arrived on the island, she was exhausted, starving, and tired of biking. To appease her, we immediately went into town for some midafternoon baked goods. I have never seen her eat so much, but riding sixty-some miles does burn calories.

Kate had recently taken a new job in a preeminent cancer research laboratory in Seattle, and she was talking about applying for medical school.

That had been her goal in college, before her dad had passed away. When she returned to school after his death, she worried that she was too far behind to be a good candidate for med school. She was always more competitive with herself than I'd realized, and her grief only seemed to exacerbate it.

As we sat there, gorging on cinnamon rolls, a man stopped by to ask us for directions. He looked about in his late sixties and was most likely retired, either living part-time on the island or visiting from Seattle, I imagined. I explained how to find the local winery between town and the ferry.

"By the way, what are you two doing here?" he asked.

"I am working on a farm on the island for the summer." I promptly replied. Kate described her laboratory job.

"We went to college together and were both biology majors," I added.

He looked at us both and then turned to Kate.

"Well, good to know you are doing something useful with your degree. I am myself a retired doctor," he said smugly.

They continued to talk while I slumped back against the bench. Why was laboratory research and medicine a better use of a biology degree than growing food? Isn't good food just as important as pills and prescriptions? As they finished their conversation, he tossed a judgmental smirk in my direction and made his way back to his car, sticky bun in hand.

We biked back to the farm and spent a leisurely afternoon touring the garden and pastures. Her parents had both been doctors, and she'd been raised in a rural part of Washington on some acreage. While she didn't have aspirations to be a farmer, she appreciated the farm life. And even though she didn't ask a lot of questions about my work on this farm, it was still nice to see a familiar face. She fell asleep early in the evening, exhausted from her long ride, and I rode with her back to the ferry the next day.

⌖

The Unsettling of America: Culture and Agriculture by Wendell Berry was on Gerry's farm library shelf. In that classic work, published in 1970, Berry methodically and artfully argues that a disconnection from land and the dismantling of small farm economies by agribusiness in this country is the root cause of our environmental degradation, food insecurity, and loss of community and spiritual fulfillment. I devoured it, reading late into the night despite our early mornings on the farm. Berry's discussion about nurturers versus exploiters, fidelity to nature, and being "in place and free" spoke to the discontent I'd experienced in my own life and reflected back in harsh light my own obvious disconnection from the land. Had my life been built around exploitation and consumption? Could I buck societal trends and learn to live my life as a nurturer? Would working my way back to the land make me happy?

Writing almost thirty years earlier, Berry had painted the picture of what I thought I wanted my life to be, and his writing helped me validate my own still-forming beliefs—the importance of sustainable food systems, a value on human culture and ecology, the possibility of a deep and serene connection with nature. No matter where we live or what we do for a living, "our bodies live by farming," he writes. I was convinced that farming was good, useful work.

⊷

While I spent most of my days working side by side with Gerry, his wife and I did not spend much time alone together over the summer, except for a week early on when Gerry was away on an educational farm tour. (He was one of the chosen delegates from a group of university faculty and farmers to participate in a learning exchange with Cuba.) Marie and I milked together and shared meals that week, and she even took me out to a small café on the island for a special birthday dinner. At that point I was still pretty new to the farm and was touched by her efforts to make me feel welcome.

On our way back from dinner, we could hear their German shepherd, Ellie, barking in a panic. As we parked the car and ran to find her, we noticed she was near the retention pond staring intensely at something in the woods. "A lamb," gasped Marie. Sure enough, a lamb had snuck through the back fence and was now being guarded by their loyal Ellie. As we crept up behind the dog, she did not know whether to greet us or keep watch on her target. We startled the lamb, who leapt high and fast behind a few trees and over to the path leading to the pond. Ellie stayed on her and picked up her trot to a full sprint. Was she going to attack the lamb? Marie had no idea what the dog would do, so we chased after her in terror. Just as it looked like she was about to grab the hind leg of the lamb, the frail babe took one more long leap and landed in the pond.

Silence, and then a few bubbles crept to the surface, followed by a shaggy white head and a bleat. It was swimming! I hadn't known lambs could swim. Marie and I both paused and then laughed for joy. Suddenly we heard a splash and I saw the wave of Ellie's tail diving into the water. I ran around the perimeter of the lake to try to call her back, but it was too late. Their chase continued underwater. I ran around to the other side of the pond and jumped in with my clothes, grabbed the lamb's tiny leg, and led it to shore. The little lamb seemed terrified but also relieved.

Like the previous summer, the days of sun, dirt, and physical exertion took their toll. Gerry and I worked well together despite the half century between us. He was overly critical at times, and the more people I met on the island the more stories I heard about past interns who weren't willing to stay beyond a few months. I was his favorite intern, he once said. His own children, who would occasionally visit the farm from Seattle, did not want to be farmers.

"It takes that kind of personality to run a farm," one CSA customer told me with a grin.

By August, Gerry had asked me to carry on as his farm manager indefinitely. According to his offer, I could continue as I was doing for room and board, and I'd have the opportunity to make my own profit running a larger version of their current CSA on their land next summer. At that point the idea of having my own farm seemed far away, and I was starting to feel the pangs of isolation. Could I really live this way? Island life offered its own beauty, but a part of me craved more involvement in the world—more people.

❧

By the end of the summer, I had set up a meeting at the WSU–Mount Vernon Research Unit to learn more about agriculture research. Wendell Berry's feeling about land-grant universities and the dangers of specialization, shared so eloquently in the *Unsettling of America*, echoed in my head. The very old, square building resembled a prison with a large metal fence surrounding it on all sides. It turned out that the scientist featured in the documentary about potatoes I had seen at the community center was based here and had funding for a graduate student. I was hesitant to go back to school but fascinated by the idea of a research greenhouse, laboratory, and farm—and getting paid to get a graduate degree. I had, after all, really enjoyed measuring those beans all summer.

One evening at milking, I just couldn't seem to get into a rhythm with Gerry. He grumbled and then started to tell me I was being too slow. In my mind, I was milking in the same reliable way I'd been milking all summer. *Why was this an issue now? What had ruffled his feathers?* Then a hoof rose up quickly and released back down, kicking the almost-full bucket of milk to the side. Gerry blamed me and tore off on an awful tirade, half in garbled English and half in German. I stood up and left the barn.

I grabbed my bike and just started riding. Looking back, I think he sensed I was slipping away, so he reacted by pushing me harder. The gray area between experiential education and exploitation is hard to navigate as an intern. Some days I still felt fairly ignorant about how to run a farm. I wasn't sure where I was going next, but I didn't feel like I had a future there. On that late summer afternoon, as I looked over the parched hillside sloping to the sea, all I could hear were waves crashing into the edge of the world.

Was I giving up on farming by looking into going back to school? That retired doctor's comments still lingered in my mind. Although I'd been offered the job of staying on Gerry's farm, I started to think practically about the low pay and lack of benefits. *Was it silly to think of farming as a profession? Was this all just a phase?* I admired Gerry and Marie's partnership and also lamented the fact that I did not have a partner of my own. I could never manage this type of lifestyle solo, I thought.

After I got a call from the admissions office telling me I had been accepted into the graduate school program with a funded assistantship, I gave Gerry my notice. Both he and Marie seemed upset about my departure, but they held back their opinions, which I appreciated.

"Get as much education as you can has been my philosophy," said Gerry half-heartedly.

He spoke with a deflated tone that revealed his disappointment. After all of those hours of work together, I was better at reading him than he realized.

The three of us had a short goodbye, and I promised to visit over the winter. It would be six years before I made it back there again. As I left, the cattle were grazing along the fence line by the road. They were indifferent to my leaving, but I looked for a response from them nonetheless. It was a Sunday afternoon, and as I sat on the edge of the ferry pulling away from the island, I began to cry. I couldn't tell if I was losing myself or just growing up. Light jetted out between cloud breaks and I felt shards of its warmth across my face. The water was glistening, and as the island got smaller and smaller behind me, I turned toward the mainland and never looked back.

Water

IN SUMMER, I WAKE UP EARLY. I have to check the irrigation on the squash. Get the order picked, boxed, labeled, and ready for delivery. I also have to make sure the crew of two (mostly inexperienced) people have enough to do. We are able to afford two part-time employees for the season. There's always weeding, but if I'm not around to watch, they tend to slow down. Finding the right help on the farm continues to be a challenge.

I'm feeling rushed and scattered today. I didn't sleep well. After the first few seasons at farmer's markets, I have decided to sell organic produce through wholesale channels—a sector that has an endless demand, it seems. I hope. I run out to the fence line to make sure the irrigation is running in the south fields. It is—the arc of water is going full force and reaching all the plants it needs to. While standing there my phone dings— another order. The Seattle wholesalers wake up early and leave early, and I need to be ready for them. I need to be grateful.

I nod into the phone, scribble the order on an old piece of paper from my back pocket. Luckily, I had a pencil behind my ear. Ten cases of baby squash by tomorrow. "Excellent, thanks so much! We've got that for you."

Baby squash is loathsome. It pays well, but it is murder on the back to bend and carefully remove the prepubescent yellow rounds hiding within the lush prickly foliage. Most of our order will end up on a cruise ship, they tell me, but I never know for sure.

I put the piece of paper back in my pocket, slip the pencil back to its place, and shove my phone into my back pocket. I wasn't planning on another order between today and tomorrow. There is already a long list of things to do and pick. I run my hand through my unwashed hair. I left my hat in the kitchen. I take a small step back and feel an unusual freedom—the lightness of flying—followed by the unforgiving panic of gravity. My arms flap like an earthbound chicken. I am falling. I have backed into an old water trough used to catch rainwater from the barn roof. I'd almost forgot it was there, and before I realize it, I'm soaked. My phone has drowned. I pull myself out and up and shake like a dog. Did anyone see? Please tell me my employees are not here yet. Thankfully, they are not.

There are days when I feel like I live in a kind of paradise, a heaven on E]arth. This small farming valley on the edge of the sea, speckled with artists and farmers and chefs and makers all celebrating the beautiful coastal ecosystem we share and inhabit together. Home of big rivers moving water from the highest glaciers to the bay, fierce eagles and slender blue herons alike scouring the delta for food, the sacred pathways of salmon.

And then there are days when I fall on my ass.

Dean is off running errands, so he didn't see me either. My plan had been to do this farm alone, and then he serendipitously arrived in my life. His tall frame and gentle brown eyes were what first drew me in. He spoke quietly and with intention. His large carpenter's hands were strong and calloused. We quickly realized that we shared a passion for music and good food and a reverence for this special valley. He didn't want to be a farmer, but he thought he could help. I think, maybe, he could see I was in over my head, and he is someone inclined to offer help.

Safe, but saturated, I feel like the ducks are laughing at me. Their quacks always make it sound like everything is hilarious. I stand in silence, staring at my black-screened phone. In shock, I wobble my way back to the house, dripping along the way, imagining what I must look like. Did I catch air? The pain in my tailbone is starting to tell me something.

My mind has always moved faster than my body, even more so on the farm. With a smartphone I can do business out in the field. It's convenient until I end up in a water trough.

At this point, three growing seasons in on the farm, I am starting to feel like I have made progress. Produce is in demand, our infrastructure is expanding. We have a homemade walk-in cooler. And yet I feel as much doubt as I did when I first moved in. Paralyzing doubt that has sent me to a meditation retreat in order to "check in" with myself. I am always "checking in" with myself, or seeking validation from outside sources—teachers and mentors, sages and astrologers—who might offer some reassurance that my divergent path is not some sort of accident or anomaly, a quarter-life crisis.

I wonder what Wendell Berry would think of our farm, of me packing up boxes of immature cucurbits to send far away on cruise ships. We are part of the machine now and still can't pay ourselves or the employees well. It seems like in future seasons we will need to hire more people and pay them less. Grow more food on more land in order to become a reliable farm with constant, consistent, blemish-free product for the insatiable wholesale organic market. The pressure to grow is formidable. And as an older farmer once told me, eventually the farmer has to "get off the tractor". to do the business work, the selling and orchestrating, and leave the farm work to underpaid labor. It is the only way to compete.

⟿

Water is a controversial topic in agriculture. More specifically, water rights, the ability to draw from the river for irrigation or build a well on your land. Instream flow rules maintain healthy water levels for river ecosystems so that they can support important species like salmon. However, farms are needing more and more water each year, especially as drought becomes the norm. Water rights have a hierarchy that often sets farm viability and conservation, including treaty rights, at odds.

This season there are strikes at a local berry farm as well. Workers are asking for more breaks and better wages. The harvesting schedule for berries is fast and furious, especially when rain is in the forecast. Rain can cause a fragile raspberry to rot on the vine.

Berries are still dirt cheap. I imagine that the farm could pay their labor force more, but then the farm might not succeed as a business. There are protests at the farm's retail market, the place where I get strawberry milkshakes each summer. What many of these young protestors don't know is how much that particular farm has done to save farmland in the valley while providing a reliable food source. The owner-family of former Japanese immigrants made their way from fieldworkers to farm owners only to be almost shut down by human rights boycotts. I can see both sides; I know now how incredibly hard it is to make a farm work economically.

Berry writes, "The growth of the exploiters' revolution on this continent has been accompanied by the growth of the idea that work is beneath human dignity, particularly any form of hand work. We have made it our overriding ambition to escape work, and as a consequence have debased work until it is only fit to escape from." After the first few farming seasons here, I wondered if the farm was working well. Was this what I wanted? This rushed and panicked feeling? The paranoia that we're just too small to succeed? Thinking back to Berry's writing, I wonder whether I am an exploiter or a nurturer.

⌖

Falling in water was a good wake-up call.

Beyond my romantic notions of taking care of the land and plants and animals, I realized I have a lot to learn about the fluidity of business, and a farm is as much a business as it is a science—and an art. Most of the picturesque family farms around us are really vertically integrated corporations. My work experiences had been in education and academics, where projects are grant-funded and slow going. I had a pseudo business plan for the farm based on beauty and fairness, but it needed a lot of work. It needed an edge.

While I had fond memories of the herb farm and of Gerry's place, my deep-dives into the workings of farms, I hadn't really gotten into the nitty-gritty of their business plans. In the case of the herb farm, the farm itself never really made money. The business made money through the global sales of the value-added products, like tinctures. And Gerry, as a retired professor, was less focused on personal income and health care (that was taken care of through his other work), and so the balance of what came on and off the farm was his focus. His farm profited, but his situation would be hard to replicate at my stage in life.

Our shift toward wholesale was loosely modeled after medium-sized vegetable farms. I decided not to do farmer's markets or have a CSA like a lot of the other small vegetable farmers in the area. Honestly, those markets were already too saturated from the popularity of the local farm movement in our region. Mixed vegetable farming is gorgeous—the gleaming carrots and kale plants—and also extremely difficult to get started.

The "get big or get out" pressure I was feeling confused me. Dean and I were working other jobs, teaching and carpentry, in the off-season to support ourselves, and I was fortunate to have had my education and early adult life subsidized by my parent's money and insurance. They had nothing from the get-go and worked hard, and I didn't take the cushion they offered me for granted. I probably wouldn't have taken on the risk of farming without it.

Dean had officially moved onto the farm a few years earlier. On the one hand, more land and people could mean that we could support ourselves from farm income alone. However, by expanding and being more efficient we would lose control over how crops were produced and harvested; we would need bigger equipment; and also, we would need to hire large crews of workers, largely undocumented, to do the work affordably.

Determined to de-hobbify this piece of land, I wanted to believe that a small farm could exist and even thrive within our complicated, global, industrialized food complex. It could be a working business and not just weekend recreation. There is nothing wrong with that, but I had something to prove. At one time this farm had been a working dairy farm, and although one of the more recent owners had refurbished the old barn into a workshop space, I still believed we would make the farm into more. A lifestyle and a way of life. Abraham Lincoln once said, "The greatest fine art of the future will be the making of a comfortable living from a small piece of land." I desperately wanted that to be true.

Skagit Valley, 2003

BEFORE I TURNED TWENTY-FIVE, I left Lopez and entered graduate school in agricultural science (plant pathology to be specific), filled with a sense of purpose and direction. My research project was to understand the late blight disease cycle in this valley and how to control it. I had never heard of a land-grant university or been in charge of my own research project, and I was excited to be in a brand-new place that seemed to hold endless possibility.

The clouds huddled overhead as I began the morning drive through the valley mist. Everything was in hiding at that hour. The sun's dull glow rose from behind the enshrouded mountains to the east. There's a retired fire lookout up there on Three Fingers Peak, and I tried to imagine the view of this soggy floodplain from that elevation. The sound of foraging snow geese directed my attention back to the late winter fields speckled with morning frost.

Tucked away into the soft folds of my scarf and coat, hunched over the steering wheel, I passed by raspberry canes, woven and strung to their wire support lines. They looked bare and relaxed, like arms stretched out on a railing. The soft green buds would soon develop, and then the berries would come in multitudes. In the next field, green daffodil heads were poised to emerge from their polyps into a sea of brilliant yellow, floral signals that illumine an imminent spring.

Raspberry and tulips are only two of ninety different types of crops, from rutabagas to spinach seed, that can be grown in the Skagit Valley. The potato, however, has become the most popular commercial crop. It still offers a fairly high return to farmers as a fresh market item. During a time when processing facilities are closing, and the middlemen, the distributors, are siphoning off food revenues, the potato remains a stronghold, a viable wholesale crop.

Most of the potato farmers here are from third- or fourth-generation Dutch or Norwegian immigrant families. Their ancestors arrived in this floodplain, recreated the dike system of their homeland, and began growing anything they could sell. As a native of the Central and South American highlands, the potato can thrive in the cool, maritime climate of western Washington. Specialty varieties like the yellow-fleshed Yukon Gold and the red Chieftain—direct descendants of native Andean tubers—do particularly well in this part of the world. The marriage of the immigrant and the potato made for a lucrative industry here, a new chapter where old world and new world could come together in a new story.

But there are problems with growing potatoes in this part of the world. The summer rains that persist into June and the lingering leaf moisture can create the perfect conditions for fungal and bacterial diseases to invade plant tissue. These minute organisms are generally foreign to most potato eaters. We hardly notice filaments in between plant cells or motile cells swimming into tuber eyes. While tasteless to our palates, these pathogens are at the forefront of every grower's mind because they can kill the potato plant so easily.

One well-known pathogen is *Phytophthora infestans*, which in translation means "plant destroyer." This causes the disease called late blight, and I have been told it keeps some of the farmers tossing and turning with fear at night. Although the blight has been studied for over a century, ever since the notorious Irish potato famine, this pathogen can still destroy a field of potatoes in a few days to weeks, leaving a farmer with a huge loss for the season in addition to a reputation as a bad farmer among local growing circles.

Only through a microscope is it possible to see the branching strands of mycelium that make up this fungus-like body. I say "fungus-like" because technically the genus *Phytophthora* falls into a whole other kingdom of organisms that are distinct from plants or fungi. At the microscopic scale, the "plant destroyer" appeared almost angelic, swaying rhythmically under the gentle pulses of my hovering breath as I peered through the scope.

How astonishing that such a simple translucent being can cause so much destruction so quickly. On the ends of each delicate branch are the luminous, lemon-shaped spores that distinguish the organism from the rest of its close relatives. *Phytophthora* can also make small motile spores called zoospores that are sperm-like in appearance and have the ability to swim through water in soil and on plant leaves. Oopsores, the circular resting spores that this organism also produces within its life cycle, will wait years in the soil for a compatible host. Then, like a patient seed, the oospores germinate, spread through the soil, and colonize potato tubers and roots to begin the cycle of destruction once again.

I remember walking out into an organic potato field. The field was in a desperate state. Although a lot of the field's foliage was still green, there was an obvious brown center of rotting leaves and stems about ten feet inward from the road. The potato plants in that part of the field had turned yellow and brown, and some were slimy and girdled, hanging over each other in a last-ditch effort to avoid collapsing. Some of the still-green leaves had the characteristic coffee-colored lesions that were surely late blight; under the leaves I could see the white mycelium, so celestial in the laboratory, flourishing among this decay.

Late blight is a sensitive subject in this area since it can spread between fields with ease; discussion over how to manage the problem can become riotous, especially the question of whether or not to use fungicides. There are prophylactic chemical options for preventing the invasion of this pathogen on the plant, which translates into weekly sprays of synthetic mixtures or basic metals, like copper—allowed under organic farming rules—that limit the leaf space available for invasion. While not applied directly to the edible part of the plant, many of the synthetic sprays are not allowed in organic systems and are toxic to the surrounding environment. There is no cure for late blight, so once it lands in an unprotected field it can run wild, making the growing of organic potatoes very difficult. Agricultural scientists have developed models to predict the appearance of this sly organism, based on temperature and relative leaf moisture, but nothing is exact enough to prevent the farmer from spraying. The risk is too high. When people say that farming is risky, they largely mean that it is sometimes near impossible to keep living things alive as a profession when nature offers such formidable opponents.

⊸

At the turn of the century, the Morrill and Hatch Acts established our land-grant universities as we know them now. The land-grant colleges were originally formed to support and educate farmers. Their aim was to encourage a diverse knowledge base in young farmers and cultivate in them a hankering to return to their family farms to do the work of farming. However, over the past twenty years these institutions have emphasized the importance of advancing the technology of agriculture. Berry writes, "The land-grant college legislation obviously calls for a system of local institutions responding to local needs and local problems. What we have instead is a system of institutions which more and more resemble one another, like airports and motels, made increasingly uniform by the transience or rootlessness of their career-oriented faculties and the consequent inability to respond to local conditions." As a former farm intern turned graduate student, I realized that I was officially turning in my dirty farm fingernails for a lab coat and sterile petri dishes.

⊸

The research center, the place I had visited while I was still interning at Gerry's and where I was now starting my research assistantship, resembled an old military base. Unlike most students, I would be spending most of my time off campus at the center rather than on the main campus in Pullman, on the east side of the state. There were few other students. We were surrounded by a hundred acres of research farmland. The soil there did not look the like the farmland I had worked over the past few years. It was gray and thin. The echoes of Seamus Heaney's famous poem "Digging" followed me:

The cold smell of potato mould, the squelch and slap
Of soggy peat, the curt cuts of an edge
Through living roots awaken in my head.
But I've no spade to follow men like them.

Between my finger and my thumb
The squat pen rests.
I'll dig with it.

At the herb farm and at Gerry's, I discovered that I loved the work of farming, but now I had chosen my own "squat pen." I hoped I would get back to being on the land someday, but more learning was the next right step and being in school again offered a strange sense of security. I didn't have to have my whole life figured out just yet.

Finding a temporary rental in a rural area was not easy—there were a few Craigslist postings and some newspaper listings. The best options were a small room overlooking the interstate (and train tracks) or a shared farmhouse upriver east of I-5. From the pictures, the farmhouse looked quaint and was on ten partially forested acres, so I called my dad and asked him to drive out there with me.

We followed a long, winding, muddy driveway to a small, blue modular house surrounded by over twenty loitering cats and a paddock of at least that many pot-bellied pigs with a replica village of downtown Winthrop (a western-themed tourist town on the other side of the mountains) made out of wood. You never know what people are going to do with their land.

The owner, a woman in her sixties, was jovial, and the house was relatively clean considering the cat population. I could tell, though, that my dad's allergies were kicking in and he didn't want to be there long. She told us that most of the rooms were rented short-term by men working at the refineries. From the strained expression on his face, I was pretty sure my dad would forbid me to stay there.

"Where the hell did you find that place, Jess?" he gasped as we sped out the driveway. We both laughed and decided to drive through some neighborhoods west of the freeway.

Eventually, we passed a small bed-and-breakfast and noticed that, like many of the other farm properties, it had a lot of outbuildings. I called, enchanted by the gardens and what looked like an old water tower that had been converted into a living space. The couple, Clark and Hilda, had four grown sons who'd left the house. I explained that I'd worked at a bed-and-breakfast in college, and I needed a short-term apartment, that I was doing research at the WSU center and then going to Pullman to take classes for a year before coming back to finish my research project.

Clark was warm and seemed almost set on the idea, but he needed to talk to his wife. He called back a few minutes later and told me that they were taking a trip in October to China. Since I had worked at a bed-and-breakfast, they were wondering if I would consider inn-sitting for the first month—and then I could stay on in the back house for the rest of the fall. I jumped at the chance, knowing instinctively that it'd be a great first place to land.

Potato and raspberry fields surrounded their home. As I watched the daily potato-digging outside my window that fall, I imagined I was Heaney. I had given up my farm life for a chance to learn and write. However, I was still having a hard time convincing myself I had made the right choice. Like Heaney, after working as an intern on two farms and feeling unsure how to ever start my own farm, I had no "spade to follow men like them"—but I had a paid opportunity to "use" my biology degree and that felt like a responsible choice.

⮵

My graduate adviser was kind, but firm. She was a woman in her early fifties, married to a veterinarian with two adult children about to leave their stable nest. She had studied and moved up through academia in a department of all men. Her mentoring philosophy was to watch me sink or swim, just as she had been mentored by her stern and successful male predecessors. The first time she told me I was doing a good job was when I was in the lab until 8:00 p.m. analyzing potato specimens. Working late was a sign I was committed. We had a general understanding I shouldn't expect time off; weekend and evening work was a clear sign of dedication regardless of what I had completed during the week.

My classes were notably lacking any ecological foundation, but they brought me into the world of agricultural science: the minutia of food plants and production. There are so many bacteria and fungi and other microbes that live on our food plants and in the soil, and decades of scientists had already documented and identified them. I had no idea! I took classes on plant diseases and soil science as well as potato science (yes, an entire semester on potatoes). The land-grant university offered a treasure trove of farming knowledge and experience, extending back from the pre-chemical era to the current molecular and genetic paradigm.

Research spanned areas of food safety, plant health, soil fertility, and plant breeding. The more I learned about the land-grant university and the extension system, the more I realized my own ignorance. I had two farming seasons under my belt, and that was all.

In Pullman, where I took my graduate classes, I found an apartment outside of town. A couple had adopted a small patch of a wheatfield that was about a forty- or sixty-acre sliver out of the thousands of acres of wheat and lentil fields around us. They were deliberately trying to rewild it. There was a pond speckled with red-winged blackbirds, rows of evergreen trees, and a huge ostentation of peacocks. My apartment was attached to the west side of their home. I had my own entrance, and one of their male peacocks made sure to greet me most mornings as I left for school, his feathers fanned. I arrived in January with snow blanketing the wheat stubble.

One day I put on my cross-country skis and took off. Once I reached the property boundary, I kept going, following the curved lines of harvested wheat. As I reached the pinnacle of one of the rounded hills, I looked out on the miles and miles of cultivated land before me. The Palouse, the wide and expansive rolling hills of southeastern Washington, was overwhelming, and I couldn't quite process the scale of food production that existed on that side of the state. Wheat is grown on thousands of acres; most of it is shipped to Asian markets for noodles.

When I entered WSU, a dramatic change in perspective was taking place among academics regarding the approaches and ethics of agricultural science. The first organic farming major in the country was being developed at that time and would be accredited before I finished. However, as in every other state, the chemical companies and large corporate farms were still deeply nested funders of agricultural colleges, so the majority of research was focused on testing chemicals and developing varieties directed toward corporate patents. With little reliable state funding, researchers needed to follow the money.

I was in school with a surprising diversity of people. When I arrived on the Palouse, I didn't know what to expect. There were a few students like me who had worked on farms and who were interested in the science behind agriculture as well as the fieldwork of agriculture. However, in my department, the majority of graduate students came from international institutions. Many students were from India or China and in some cases South American countries like Uruguay. While the department tried to facilitate social activities, I was on the main campus for only two semesters and never got a chance to know my peers well.

When I was on campus, I often found it hard to connect with other students. They worked eighty to a hundred hours a week in their labs and did not take time to socialize. I remember a Chinese student who scheduled her C-section for spring break in order to be back in the lab for spring quarter. I wasn't sure if she did this on her own or if her adviser set her schedule. Either way, there was no maternity leave, and she didn't seem to mind; she just kept working.

When I returned to Mount Vernon to do research for the summer, the college had renovated an old farmhouse adjacent to the research center where students could live. They were at the start of a major renovation that would eventually include a huge fundraising campaign and a new research facility to replace the old, decrepit one. That summer I was a guinea pig in the new housing option. I chose a small bedroom on the southwest side of the house. A large black locust tree, planted by the family who'd settled that piece of farmland and then later donated it to the college, tapped my window. Another graduate student, Diego, lived downstairs with his wife and son. Renee, an entomologist from Canada, also stayed at the house during the week when she was too tired to commute the hour or so back home over the border.

Diego's wife, Laurie, was home all day, so she took command of our shared kitchen. She cooked all Diego's meals and did all his laundry. When we wandered back to the house after long days of field trials, I found myself mixing a box of macaroni and cheese while Diego had a four-course meal and a clean stack of clothes waiting for him.

"Isn't it nice to be Diego?" Renee once said with a snicker.

One day as I was passing between my office and the lab, I noticed an attractive, tall farmer standing in the doorway. I assumed he was a farmer because he was wearing brown rubber boots. His hair was mousey brown and tucked up under a hipster foam dome hat.

"Hi, can I help you?" I asked.

"Hi, my name is Jim. What is this? I found a bunch all over my truck this morning."

He stuck out his long slender finger and showed me a brown and orange fuzzy caterpillar.

"I think that will eventually be a moth, maybe one of those yellow butterflies you see around." My heart fluttered like a butterfly, but I tried to play it cool. "I can rear it if you want me to. It's no problem."

He tilted his head in confusion and slightly smirked.

"I mean, my friend studies insects and we can raise it in her lab to see what it becomes and then we can identify it." I shuddered at the sheer nerdiness of my response.

He eyes perked up with curiosity and still, perhaps, a little amusement.

"Great, you can take it and give me a call when you find out. I figured this was something you people do here."

He handed me the creature, and I quickly scrambled back to the office to grab my lunch container. As I scurried back from the office, he handed me a crumpled piece of paper with a phone number written on it.

"Do you live nearby?" I asked.

"Yeah, I am farm manager on a local organic farm south of here."

"I would love to see your farm sometime," I said, shocked that I could be so forward but excited to finally meet an organic farmer after working with so many large, conventional potato farms.

He said "sure" in a fairly noncommittal way.

"I hope to find out who this little guy is—we want to make sure it isn't going to eat our product, you know."

With that he nodded and headed back out the door, never asking me my name. I smiled and scurried to the entomology lab to tell my insect friend what happened.

Renee was one of the few people my age in the valley, and we became fast friends, bingeing on pizza and old *Sex and the City* seasons at least once a week after we finished our work. She worked longer hours than me and also managed this long commute.

I ran into her lab, reciting the events of the past five minutes very rapidly. She smiled and gently pulled out one of her special plastic boxes with the holes on top.

"Let's see what this beauty will be," she said with a wink.

I blushed again and watched her load the caterpillar into the box.

"We will see," I thought to myself gleefully.

After a few weeks, the caterpillar died. Renee apologized for forgetting to ask me to feed it over a hot weekend when she was back in Canada. I told her not to worry since I still had a reason to call that farmer back. I fished his number out of my office drawer and we chatted over the phone. He invited me to see the farm and I jumped at the chance. That visit would turn out to be one of many excursions away from the research center. The farm, only a few miles away, would soon feel like a new sort of home.

⊕

When I returned to Mount Vernon from Pullman for good after finishing my classwork, I searched for housing again. It was January and I would be there for at least an entire calendar year, so I wanted something more private than Clark's bed-and-breakfast or the student farmhouse. From a posting at the local co-op, I found a small one-person cabin adjacent to a Native American reservation. Focused on the farming community and still learning the history of the area, I didn't know very much about the valley's long, complicated history of colonization.

The Swinomish are a Lushootseed-speaking coastal tribe with about a thousand enrolled members living on or near their fifteen-square-mile reservation, created by the 1855 Treaty of Point Elliott, located between the Skagit Valley and Fidalgo Island. They manage 2,900 acres of tideland and 7,450 acres of inland terrain, and the tribe owns a fishery and canning business as well as a hotel, casino, RV park, and golf course.

While the tribe also leased land to several non-Native homes, my cabin was tucked away on some of the privately owned property. The owner's house was farther back in the woods. Between the house and the cabin was their large and productive garden—built in an opening in the forest where a plane had crashed several years before. Occasionally pieces of metal from the plane's body would still turn up in the soil as they gardened.

One of my first mornings, the pipes, exposed under the raised cabin, froze; I called my new landlords. A few minutes later I saw them puffing down the path, wearing puffy coats, pajama pants, and large wool caps. One of them held a hair dryer.

"Do you have a toolbox?" the husband asked me.

I pulled out a shiny, brand-new tool kit my mom had bought me from Home Depot.

"Well, that's fancy," the husband chuckled.

It was filled with tools I had no idea how to use. I handed it to him and stepped back.

After about twenty minutes of blowing the pipes with the hair dryer and monkeying with various parts of their homemade plumbing system, the pipes started to thaw. This would become one of many new skills I would learn living alone.

The cottage called for a cat, partly for the mice and partly for companionship. I was starting to enjoy being by myself, but I still did not feel content. I went to the Humane Society and spent hours touring aisles of various pit bull crosses—a sure sign of living in a rural area. In the West, meth production is common in many small towns. Most operations have some sort of pit-bull-like dogs to guard the door. Mostly they aren't fixed, and so at the shelter I saw the most bizarre blends, from a pit-chihuahua to a pit-poodle.

But I knew I needed a cat. Growing up, my dad had been allergic to cats, so we never had one after one brief experiment. My first memory was watching a tiny black kitten march down the hall and jump on my parents' bed, nestling atop my dad's head. The kitten was gone pretty quickly after that. We also had a dog growing up, but we stayed a one-dog family. That was enough animal for the four of us, according to my mother.

Jeffrey was a large gray cat. When I found him huddled in the corner of his cage, he let out a loud "meh," which I soon learned was his signature sound. He was slow to come over and let me pet him, but once he did, we bonded. I saw something bold and loving in his eyes. At the cabin he was my protector, often sitting on the roof over the front door when I came home, like a proud lion. He did bring me lots of prizes—a squirrelly tail, various organs, and once, a mouse heart, which he carefully laid out on my pillow.

One night I was startled out of my chair by a large boom. I worried at first that another plane had crashed in the garden. Then there was the rapid fire of small booms. Bullets? I grabbed Jeffrey and tried to decide what to do. Together we decided to call the police.

When they arrived, I slowly opened the door. I wasn't used to getting visitors there.

"Wow, that is a fat cat!" the officer exclaimed.

"His name is Jeffrey," I responded.

They went on to explain that I was technically living on a reservation, and the tribe liked to light off fireworks at various events. That was what I was hearing.

"There isn't much we can do, ma'am."

Once I knew it wasn't some crazy person firing off in the woods, I was fine. Later that year, driving home on the Fourth of July, I had a taste of what active warfare feels like. I had to drive through the main village on the reservation to get to town. Fireworks shot over my car and alongside my window. All I could do was speed forward and hold the wheel tightly. It was dazzling, the light and the sparkles, like I was descending from space.

On the days when lab work seemed overwhelming, I would drift off to Jim's farm to see what was happening. The owner, Henry, was kind and charismatic, and I was grateful that he let me hang around while Jim was working. It was the largest and oldest organic enterprise in the valley, and they grew crops in several areas around the Skagit River. It was my oasis; the leek, potato, and chard fields stretched out to the river's edge. We would walk through acres of organic vegetable fields and watch the blazing orange sun sink behind Mount Erie. Jim's hours were unpredictable and generally matched my unending days in the lab. On Sundays, the farm was always closed. We would go to our favorite dingy diner for brunch, where the sprite waitress would always ask, "Bacon crispy?" Afterward we would take a walk or escape to the nearby Cascades to wander in their serene beauty. It was a routine that I both looked forward to and relied on after each long week.

Despite having had various boyfriends on and off over the years since college, some more serious than others, I could tell something was different about this budding relationship with Jim. My experience with him was so intertwined with discovery in these fields and mountains, the valley and he became one. It started off slowly and soon became a full-fledged relationship with daily calls and plans, and secret kisses behind the barn. We were together most weekends. Our connection was young and playful and I finally had someone to share all this beauty with—the piercing sunsets, the trumpeter swans resting in the fields, the cold, swift river.

❧

In my second and final year of research, I struggled to complete my project. My adviser had a certain intensity made more extreme by her new role as interim director of the research station and its mammoth renovation project. Feeling stifled, I needed to fight for control of my own time again. I tried to please her, and I was always ready to take direction, but I felt like I was never good enough and could misstep at any time. There was pressure to finish my project and move on, but I was still not sure where I was going.

Our research had discovered that the pathogen most likely did not survive on overwintered potatoes in the soil, but on potatoes infected with the disease in storage. As the unsold potatoes were cleaned out of the storage shed into the warmer temperatures of springtime, the pathogen was released back into the fields. That is one of the reasons the disease had persisted there.

Although I enjoyed the process of learning about late blight and how large potato farms worked, I constantly kept wondering what I wanted to do next. I kept wondering whether all this science was really helping the farmers, many of whom were older and close to retirement. It had taken me two years of rigorous focused attention to make what ultimately felt like a simple, intuitive observation. Common knowledge.

The Master Gardener program, part of the county and land-grant extension programming, had created a large garden next to the research station. It was a beautiful, sprawling sanctuary managed by volunteers. I enjoyed hiding away in the various nooks and coves created by hedgerows of roses and arbors of willows. Renewed by the garden, I could go back to the lab to keep working.

Life at a land-grant institution was fascinating. Because I was at a research and extension center, there were all sorts of extension presentations for farmers and community education events I was able to attend as a student. "Extension" largely means community and professional education—extending all of the high-level information generated by the experts doing research to the community through talks, bulletins, and websites. The focus is on translating research into practical and useful skills. People came to the center to learn how to prune apples and grape vines, make hard cider, manage weeds, and take care of soil. I soaked it all up!

As a graduate student, I volunteered to give free presentations and set up educational events on compost and plant health or whatever was holding my attention at the time. This was part of the work that I loved—the teaching and gathering of people together.

At the research center, as opposed to the large main campus university on the other side of the state, researchers had a direct relationship with the grower community. Farmers would drop off a sick plant for diagnosis, and then talk with a researcher about their options and questions. The uncertainty of farming, which is constant no matter how long your family has been doing it, was in part mitigated by that support system.

As I woke up at Jim's cabin on the edge of the farm where he worked and lived, I peered out to the fields and watched the early trailers full of vegetables head to the barn while I drank my cup of coffee. The morning light glistened on the newly emerged purplish-green kale leaves in the field. Farming was a way of life that had its own rhythm and challenges, its own freedoms and confinements.

Jim's farm was always especially busy around Thanksgiving. One night we were supposed to see a movie in town and I didn't hear from him. It was November and dark already, so surely he was home. I finished my work, and instead of going home to change I went by the farm. I was always tracking him down in one way or another, accepting that farm life was much more demanding and unpredictable than my work schedule. At that point I was writing up my results, so most of my days were spent inside typing, and I was antsy.

The farm was not one continuous stretch of land, but a patchwork of owned and leased pieces. Land sharing made it possible for mid- to large-sized farms to work economically in the valley. Jim's farm hired about a hundred temporary workers in the summer and employed about half that number during the busy holiday season in order to harvest and sell winter vegetables like leeks.

I saw the farm truck from the road, and I found Jim in one of the back fields. He was finagling a set of stadium lights on a trailer. The harvest needed to continue, and the harvesters needed light, so this was the solution. As I parked my car and walked down the muddy road, I saw clouds of breath in cold night sky, and I heard the rhythmic pulse of swiping machetes. The crew of about fifteen Hispanic and Indigenous workers was plowing through the field, grabbing leeks and slicing the tops off in fluid motions. It was magic to watch, that speed and efficiency, and it was also heartbreaking. It was almost freezing out and this field had no end in sight.

Jim saw me coming down the road.

"Hey, sorry, my phone died, and we had to get these lights up."

I smiled, feeling embarrassed that I was just watching and not helping.

"Wow, you guys are busy. It looks like a late night."

Jim walked toward me under the massive lights.

"Yep, a late order came in and we'll be here until at least 11:00. I am going to get these guys some pizza. Wanna come?"

I told him I was tired and I would check in tomorrow. A heaviness came over me as I drove back to my cottage, warm and safe in my car.

Did I have the stamina to do this grueling physical work?

⊷

My two-year program was almost over. A professor I greatly admired had asked me to apply to a PhD project in her program at Oregon State University. This offer was both an honor and a disaster for me. Why would I leave this place? Why would I leave Jim? I waited until the end of the summer to decide, avoiding any commitment until the last minute. It was the perfect opportunity, and I finally accepted and made plans to move to Corvallis for fall quarter. We had only a few weeks of summer left to savor.

Jim was extreme and in turn brought out the extreme in me. He took risks. He operated on little sleep, which always amazed me. We could talk for hours. He worked on a farm. I enjoyed chasing after him up backcountry ski slopes or steep trails. He never waited for me.

We grew closer each day, but I still felt too young to be settled anywhere, and the two of us were not anywhere close to getting married. I was chasing behind him in many ways, not just in the mountains, trying to keep up with his constant motion. As I debated over my possible move to Oregon, he seemed surprisingly open to the idea of a long-distance relationship. In my heart, I secretly hoped for a shared future, perhaps settling together in this special place. Inside, I desperately wanted him to have the same dream too.

Jim's parents were both chaplains. Time with his family was refreshing, and it distracted me from the fallout of my own parents' divorce and the initial years they spent reestablishing themselves. My more recent memories of my own family's holidays included disorienting movement between my parents' houses, talking about what had gone wrong and what needed to be fixed in our collective lives. Jim's family, on the other hand, always had a lively and sometimes raucous plan in the works. On Christmas Eve they dipped into their yearly fondue ritual and laughed for hours with each other around the living room. Before the holiday night was over, the family gathered around Jim's father as he read the story of Jesus's birth from the Bible. Their house was a wholesome refuge, and while I first chuckled at the Bible-reading part, a sense of calm and connection came over me that was sorely missing in my life.

⌖

Pasture has its own grace.

You look out on the late August light reflecting off the rich, verdant mix of grass and clover. You are looking at a place where you want to stay forever. You didn't grow up in the Skagit Valley, but you can't imagine living anywhere else.

You are in the kitchen. Boiling fresh, red beets. Mixing salad. Cutting bread. Shuffling Jim's miscellaneous papers and crosswords around on the table to make room for dinner. You've put on a fresh shirt, but still look disheveled from a long day of research. Your long hair always frizzes with the late summer humidity. You want to look nice. In fact, you'd bought a new dress but had rushed out of your cabin too fast to remember it that morning. You rushed a lot in those days, doing research, taking classes, getting in a few runs, and making time for him.

The phone rings. Your hands are full, and it's his house anyway (even though you are here all the time). The caller doesn't leave a message. It rings again, and again no message.

It's time to blend the soup, so you transfer the boiled beets and onions to the blender. Jim is late, but that's okay since the dinner isn't ready anyway. Who is calling? Why aren't they leaving a message? You forget about the blender, and soon the lid loosens and the kitchen instantly becomes a bloody crime scene.

You are making a special dinner because you're leaving and you have been dating this person for over a year. Your plan is to finish graduate school in Oregon and then immediately come back. You've gone back and forth on this decision, keeping the professor in limbo (and somewhat irritated), waiting until the last possible minute to accept the assistantship. You want to start a real life with Jim as soon as possible. The farmer he works for feels almost like a mentor to you, willing to answer so many of your pesky questions.

You move what is left of the pureed beets to a pan. You add salt and some pepper and stir. You'll clean the rest up later.

You know you love him, but you're never quite sure what he wants. He's always reluctant to plan, and you are a planner. Maybe you will start your own farm? You are already creating a life together, you think, in this beautiful valley.

He rushes in the door. The phone rings again.

"Don't answer it," he snaps at you.

It rings and rings, and in the silence between rings all you hear is his breath.

"We need to talk."

He doesn't see the red flesh splattered everywhere. You wipe your hands on a towel, looking at him in disbelief. He goes into the living room and sits on the couch. You reluctantly follow behind him. This was not how the night was supposed to go. This is not what you had planned.

He doesn't have to say anything. The weight of his head in his hands says enough. And maybe you'd known. There was something off, something that down in the dank grotto of your heart you never trusted.

The soup smells of burn. He tells you affairs are more common than you realize—his twisted version of consolation. Why had you tethered all your hopes and dreams, your vision of a life of farming, to him?

You are devastated, and definitely leaving.

Fire

HERE, WE BURN THINGS A LOT. Not our house trash (like some of our rural neighbors), but a lot of the cardboard and feed bags and scrap wood that builds up on the farm. The leftover drip lines, plastic mulch, and cracked plastic planting trays go to the dump. The bar on the way there will give you a free shot if you show them your dump receipt on the way home.

After one of the driest summers on record in Washington State, we finally have rain. It has replenished our dusty fields and helped to gradually tame the raging wildfires in the Cascade Mountains to the east. Berry writes, "The farm can only exist within the mystery of wilderness and natural force." We respect the forest, alpine, and nearshore ecologies around us . . . and I am still terrified of their unbridled power and potential for destruction. We are due for a massive earthquake.

Usually we experience "June-uary" in our northwestern part of the state, a prolonged period of springlike wet weather and cool temperatures, which keeps summer at bay in these parts until at least July 4. In a normal season we don't have to worry about watering our plants until late July or early August. Mother Nature offers us an ample and free water supply that germinates our seeds and bequeaths a verdant glow across our rows of freshly transplanted vegetables and herbs.

However, this year was different. Way different. Despite cunning and early efforts to upgrade our irrigation system, our usually fertile and saturated soil went bone-dry by May.

I have a fire in my belly—I am pregnant with my first child, and although my due date is not until July 9th, I've been convinced I'd give birth by the summer solstice. Admittedly, the record-high temperatures and general discomfort and hormonal shifts associated with being nine months' pregnant might have influenced my delusions. As I wait in the house, too swollen and hot to be useful on the farm, I am ready to welcome home our little girl as soon as possible.

⊸

After living together for several years, Dean and I had decided to get married. Sitting at our favorite wine bar on the nearby slough, I had blurted my own crude rendition of a proposal—now or never. My mom's favorite adages (she loved using adages as parenting tools)—"Shit or get off the pot" and "Why buy the cow if you can get the milk for free"—echoed in my head. In truth the time was right, and we'd gotten married on the day after Valentine's Day.

The following October, on a chilly Friday night, I was outside in the waning daylight packing a radicchio order. A typical fall windstorm was slowly picking up momentum. As I assembled the last box, I heard a large pop, similar to a gunshot, although duck hunting season hadn't started yet. Dean and I both ran to the house to check. Our power was out, so we decided to call it a day, have a drink and a romantic candlelight dinner. The next morning we'd find out that ours was the only house on the street to lose power; a fuse had blown at our transformer along the road. We had been too busy to notice.

⊸

Now a baby was on the way, due smack dab in the middle of a hot, busy summer. As the long and generous days unfolded, I followed the clock closely. Time moved slowly, and there was nothing I could do about it. In the hot, stagnant air I found myself stranded on the living room couch, immobile, feet elevated to reduce swelling.

I am a person who's used to doing things. Many things at once. While I consider myself a fairly observant and compassionate person, I know I can become hard and removed when the list of things to do stretches too long. Caught up in the doing, especially during farming season, I look solely at the future steps with no time to admire the ground below me. This summer, however, I had hours on end to stare at my oversized, barrel-shaped feet and wait. We had hired more help and pulled back on some of our orders to accommodate this new life expanding within me.

⚶

We got her number from a farmer friend. She came with a gallon of water. She wore a long, white, button-up shirt and jeans. Her long black hair was tied up and covered with a handkerchief. She was prepared to be in the field all day. She spoke some English and I spoke some Spanish, so we figured out how to communicate. Her full name was Maria of Guadalupe, but she said I could call her Maria.

I gave her the tour we offered to all our part-time employees. She seemed surprised by the time we took walking around and talking. She wanted to jump in and start working. She felt guilty and wanted to cut our new employee orientation (mandated by the state for all new workers) short.

"¿Qué puede hacer por tú, Jessica?"

I assumed she hadn't worked anywhere that went through these formalities. She probably had never had a legitimate job in the United States, one with workers' comp and equal rights. Agricultural workers are allowed to slip through all the cracks. She was all about getting to work.

Maria didn't take lunch breaks either, or if she did, they were fast and in her car. I imagine she was used to eating on the roadside, quickly, when she worked for the larger berry farms. She'd bring a bag of steaming tamales or some carrots and fresh fruit.

While I could still bend over, we worked together in the field. Side by side sometimes. Maria was faster than me and I struggled to keep up with her. She had just the speed and extra-skilled hands we needed. In fact, she was a master. Maria could weed, bundle herbs, and harvest vegetables faster than any employee we'd ever had. She arrived on time and was ready to get to work every day. No complaints or demands.

As we spent more time together, I was able to ask her questions, in my broken Spanish, about her life. She had four kids. Some days she went to work at the berry farm before coming to our farm. Her husband worked at a potato farm, so most days her kids were home, loosely watched by neighbors. The oldest was fourteen and had a cell phone.

❧

Despite being as big as a house, I tried to remember that patience is a virtue. I fought the waiting, ablaze with questions. Would our tender kohlrabi survive this drought? Will she have brown hair like my husband and me? Will there be rain in the ten-day forecast? Will she love music like we do? Each new day arrived with much anticipation and, as the ground remained dry and she stayed hidden in my belly, I tortured myself with unanswerable questions. I came up with new questions and worries, and in the interim I forgot to step outside and admire the golden sun as it sank behind the field. I overlooked the botanical milestones, the first vibrant squash blossoms and the aromatic unfolding of basil leaves, that usually grab my eye and inspire me through the growing season.

However, as I accepted my expectant waiting, the questions eventually slowed. The baby would arrive when she was ready. The rain would return. I had to surrender and admit that I had no control of either event. I just had to wait. My mind quieted down, and within that silence I could begin to feel the summer beauty around me, filling my whole round body.

Farming is a creative act. I was still giving birth to this farm, as I was about to give birth to my very own daughter. My flesh transformed into hers, my body a vessel. The urge to make, to make myself anew, had been with me a long time. In fact, it is how I landed on this land. Now, I was waiting for a living, breathing light to emerge. All I had to do was push.

The waiting also equaled slowing down. Since stepping onto that herb farm more than ten years before, every summer for me had been about farming and growing food. Farming meant giving up mobility and summer plans and vacations. I opted to stay and grow and often pushed myself too hard. Forging a new lifestyle and set of values for myself had come with costs, exhaustion being one of them. Now, my body wasn't my own, and I had to pay attention to every single move.

By midsummer, I had to delegate all my farm chores to Maria and Dean. However, I did spend some long evenings in the basil tunnel after temperatures had cooled down. Under this plastic enclosure, the basil thrives in the moist heat and can survive past the low evening temperatures in September. During the summer, especially that one, it is almost unbearable in there between 11:00 a.m. and 4:00 p.m. After dinner, I'd lug myself out to weed and harvest some of the luscious green leaves. The smell, when you open the sliding door, is blissful—sweet, spicy notes on the inhale, and rich, earthy relief on the exhale. Bending over was not easy, but I made it work, or, when my back ached, I'd drop to my knees. Rumi says, "Let the beauty we love be what we do. There are hundreds of ways to kneel and kiss the ground." This was one of them.

A few days before the baby was due to arrive, I got a strange call from Maria. She was crying. She said she couldn't come in.

"No problem," I said. "Are you okay?"

"Sí, sí, Jessica. Hasta mañana."

I was worried; there was a lot to do on the farm. The next day we didn't see Maria, nor the next. At the end of the week, I got a strange text from her saying that she'd had to go back to work at the berry farm full time. She was sorry.

We didn't know what to do. There wasn't anything to do. With the baby coming we needed her help. More importantly, I was worried about her. Perhaps the berry farm was able to offer her more money, but was there something else going on? As we'd become friends, she told me more about her family and life as a mother, which I'd found comforting in my pregnant state. She also shared, without going into detail, some horrible memories from the berry fields. Stories of mean bosses and competitive coworkers and, since she was working under the table, no state protections. Perhaps that is how she found her way to our farm.

As we inched closer to the delivery day, I let my hopes of Maria returning go. She didn't call me back, and I didn't know where she lived. Occasionally as I passed a mash of cars parked at the edge of a raspberry or blueberry field, I would try to spot her in the sea of hands and hats and rubber coveralls. She might have been out there, but to me she was gone.

Willamette Valley, 2006

THE DRIVE TO CORVALLIS from the Skagit Valley is about seven hours without traffic and with a few pit stops. Fran, my loyal golden lab mix, was in my rearview mirror. I hadn't been planning to get a dog, as a transient graduate student and a perpetual renter, but Fran was part of an accidental litter of eight. On the farm where Jim worked, Henry's son was going to breed Fran's mom, a golden retriever, and sell the puppies to put money toward his school fund. However, a roaming English lab from the tulip farm down the road snuck into the garage, leaving evidence of his sinister act—muddy paw prints on a well-groomed, golden back. He had traipsed across the fields to do his dirty work and the result was a batch of over-the-top, adorable, roly-poly puppies.

Despite my intention to avoid seeing or playing with the pups, I ultimately became smitten with the smallest, yellowest ball in the bunch, and, as one by one they were sold off, I had to stake my claim. Now that I was in Oregon, in a new strange town, heartbroken, searching, and distraught, I looked at her in my rearview mirror and felt a deep sense of comfort from her loyal, unfailing smile. She was a true companion.

This long drive left me with a lot of time to think and process. After the scurry and rush of packing and moving, all I had to do was sit. Why was I so devastated?

My life had been in forward motion, always hurtling toward the next ex-
perience or relationship. I hadn't taken much time to stop and think. Now
I was heartbroken and fleeing the place and person I'd thought I loved.
My life had been just beginning to make sense, and now it was a hot mess.
Why was I going school again? Where did this unsatiable interest in agri-
culture come from anyway? Where would it end?

I was raised to be tough, to soldier ahead. That is how my own mother
dealt with pain, loss, and her recent divorce. No therapists, no self-help
books. Just grit and vodka. That is what she was taught by her parents.
That was the culture of academia, too. Never show weakness, always stay
on the offense. Sink or swim. I was slowly realizing how I had wholeheart-
edly accepted that ethos, especially as a woman working in science and
agriculture. Stoicism was my strength.

The Willamette Valley in early September, I soon realized, was noticeably hotter than the mild, marine climate of the Skagit Valley. I'd spent
the day in Corvallis responding to housing ads, and within a few hours I'd
found a small apartment that accepted dogs. School would be in session
soon, so most of the landlords were eager to fill their remaining vacan-
cies. Like in most college towns, there was a wide range of options, from a
dingy back room to a brand-new, two-bedroom townhouse. As a doctoral
student I didn't want to live in a dive, but I couldn't really afford to buy
anything either.

I was in a daze as I drove through the maze of frat houses and neighborhoods lined with ample bike lanes. Trying to not remember the ground-
swell of tears from the day and night before, I focused on housing. I reminded myself that this had been my plan all along, to come down here
and continue in school. The only reason I had to go back to the valley now
was to get my furniture.

I called my mother several times along the drive south, and, as an attentive mom, she could sense that something was wrong. We have always had
a good relationship. We never really talked about deep feelings, but she
was always there. A chronic worrier, my mom was most happy doing routine things—inspecting her mole traps in the backyard or organizing her

various collections: pins, trains, coins. She also maintains a large network of friends who would do anything for her. That is something I've always envied about her.

As I explained my plan (drive down, follow up on housing ads, ideally sign a lease, stay in a cheap motel, and return the next day), I could feel her inquiring eyes boring through the phone. I could not bring myself to tell her what had happened yet. It was too raw and I still didn't have all the details. The shame and confusion I was feeling was dense and heavy and foreign to me.

My mom's voice softened into the phone, her polite way of showing that she knew something was off with me. I wasn't fooling her and I knew it.

"Why don't you visit your aunt on the coast? That might be nicer than staying in a hotel. I am sure she wouldn't mind, and it's only another hour out of your way. Maybe stay an extra night, enjoy the beaches."

I knew she was right, and I also knew that I was not in a rush to get back to the valley. I nodded silently into the phone, and she said she would give her sister a call for me.

⊷

As I curved down the cool, damp two-lane road, I could smell salt in the air. Douglas-firs and cedars draped over the road, covering my little car like a blanket. The road between Corvallis and the coast is not long, but its bends and ridges distort time and space. It is not a linear drive, and when you finally emerge at the expansiveness of beaches, you feel as if you are emerging from a damp and tender womb. Years later they would build a wider, straighter road to Newport, but I always remember the wild, windy path to Waldport.

As I became attuned to the smattering of roadside crab signs and whirly-gigs, I knew I wasn't far. The ocean greeted me head-on as I arrived at the grand expanse of water. At a turn left, south of the tourist village, are some of the most open and, at a time when I was grasping for answers, redemptive swaths of beach on the Oregon coast. It was almost sunset, and the late summer crowds were still lingering in small clusters as I edged southward. Kite skates, blankets, and hand-holding walkers were all in view beneath the sandy dunes that line that thin stretch of Highway 101.

Dee, formally named Helen after my grandmother, was my mom's youngest sibling of five. She was one of my two "hippie aunts," as we lovingly referred to them. My mom was close to her two sisters, but they had chosen notably different paths in life. Kate, the next oldest, had been married and then divorced and had pursued a long and successful career as a professional musician before retiring on the other side of the mountains in Oregon.

Dee had also married young and lived on a farm in Wisconsin. I remember visiting her cows and enormous vegetable garden when I was young. One wall was decorated with copper molds and a woodstove that heated the house with a rustic warmth. That chapter of her life eventually ran its course during her thirties, and she joined my other aunt in moving to Minneapolis and changing their last name to my grandmother's maiden name. She loved plants and worked in nurseries and flower shops before relocating to Oregon. She was someone who'd fanned my early sparks of interest in plants and farming.

Now Dee managed an art gallery and lived in a small house off Highway 101 behind the gallery. Her life, it always seemed to me, was of her creation. Each new chapter had its own original story, and she made choices in her life thoughtfully and deliberately. While I don't blame my mother for being a devoted wife (I was grateful she was always there for me and my sister) I wondered what she would have done professionally if she hadn't been a full-time mom.

I pulled into Dee's driveway slowly. I had stopped at the campground parking lot about a mile outside of town to have one more good cry as the sun set into the sea. I knew she could pry out what was bothering me in no time if she wanted to, but for that moment I wanted to have my feelings inside the security and privacy of my own car, dog as my only witness.

As I entered the driveway, I saw a soft, orange glow emanating from her windows. She was at the stove and a vibrant steam smothered her face. I walked into the smell of fresh crab, rice, and a gleaming green salad undoubtedly harvested from her own garden. The table was set, and she handed me a full glass of wine. Dee didn't have children of her own, although she had tried many years ago. She had always been a strong presence in my life and I appreciated her innate way of nurturing people. She would eventually move back to Wisconsin to care for my grandmother in her last years of life.

I didn't have to talk about anything specific, I just had to sit at the table with its beautiful spread of food. I hadn't eaten anything all day, so I inhaled the healing meal and soon curled up in her guest room under a quilt with Fran by my side.

⊕

I missed Jim as much as I hated him. We'd tried to talk on the phone during my drive, and he had offered to help me move my stuff down to Oregon after I found a place, but I did not know what to do at that point. During the sleepless night before I left, I'd gathered up all the gifts he had given me and left them in a heap at his doorstep—a dramatic gesture that offered only a brief reprieve from my pain. It looked much better in the movies. Almost immediately, I felt petty—and also lamented the loss of that beautiful spinning wheel he'd given me as a birthday present.

It didn't take long to lose my illusions of Jim completely. I guess I had known it all along. If this was what he really wanted—a farming life, a life together with me—we would be looking for land and making a plan for the future together.

Maybe I didn't need anyone else? Maybe I could go it alone? What was I trying to do, anyway? At this point I wasn't sure. I was in Oregon.

I spent the day walking the beaches with my aunt. We stayed on the hard, saturated tide line. Our footprints left temporary impressions on the wet sand. As the surf crept up to our shoes, we scuttled inland. Fran, on the other hand, ran free. Head-on into waves, sprinting back and forth between the safety of us and the curiosity of the sea, she was exuberant. Silence was interspersed with chatting, underscored by the reliable crash of a new wave.

I thought about whether I would ever see Jim's family again. I didn't know if his parents knew any details at this point. The other woman was married and had been a loyal sheep in his father's congregational flock. I wanted more information, but I savored the safety of my exile: the comfort of my aunt's company and the vast beach.

I spent another night on the coast with my aunt, avoiding all conversation about Jim, before heading out. She knew me too well to ask any probing questions. Her caring presence saved me from completely falling apart.

⌁

Although my heart ached for the wide expanses of farmland in the Skagit Valley, I became intrigued by the rolling sheep pastures and grass seed farms that line the roads from I-5 west to Corvallis, the expansive midbelly of the Willamette Valley. Knowing my aunt's refuge was only an hour away, I moved into the small purple house in Corvallis I'd rented. It was on the corner in a neighborhood south of campus. The house had a large fig tree in the backyard and a night blooming jasmine that smelled like heaven.

During my orientation to the department, I visited the research farm and learned about all of the facilities that were available to us on campus. It was refreshing to be in a college atmosphere again, and the town (Beaver Nation!) was abuzz with fall air, football, and the boon of the return of students. All the maple and oak trees lining the campus walkways turned mesmerizing colors, red and golden, and the cool crisp days were perfect for wandering in the woods around campus. It was the northwest, but it was not as wet as the Skagit Valley and not as hot and dry as Ashland.

Jane, a strong and confident woman, was my main adviser; she was in the horticulture department. She had just received tenure and was a well-known advocate for soil health and sustainable agriculture. I had followed her research and extension projects for several years and was looking forward to working with her. A robust activist-type in her early fifties, she had a lioness laugh and an intrepid way of diving in and leading a conversation.

We were going to be studying a biocontrol agent, a fungus, and its ability to control bean mold. Unlike the Skagit Valley of Washington, the Willamette Valley in Oregon is known for its processed and frozen vegetable industry, and beans are a major part of this sector of agriculture.

Historically, control of white mold had been accomplished with chemical fungicides, but as more growers wanted to become sustainable, they were turning toward nonchemical plant protection. That included the application of beneficial organisms that can colonize and destroy dangerous pathogens before they affect plant health. In the case of white mold (a fungus), a new biological product—a beneficial fungus—was on the market. My work was to understand how to use it and how the two organisms interacted—and make sure it wasn't another jug of snake oil.

Jane invited Carl, a well-known and longtime full professor in the botany and plant pathology department, to work with me on this project as a co-adviser, since he had expertise in the field of biocontrol from his many years working with apple diseases. He was a quiet man with a sweet smile who reminded me of Kevin Spacey. In comparison to my master's adviser, both Jane and Carl were easygoing; they trusted me to manage my own time and the details of my project. Freedom felt good, and I was ready to dive into this work with no distractions.

❧

The agricultural life sciences building, where Jane's lab was located, was built in 1992 and housed the horticulture and several other departments and offices, like biochemistry and soils. It was connected by a sky bridge to the old building (built in 1956) where Carl's department was located. I spent a lot of time going back and forth, and one rainy afternoon I was excited to pass a familiar face.

Abby had been a student with me in Pullman. We met when she was just graduating and about to hike the Washington part of the Pacific Crest Trail, which spans 2,650 miles from Mexico to Canada, with her boyfriend at the time. That same summer, I was returning to Mount Vernon to do my research. I never heard from her again but hoped she would turn up after she finished that long and grueling hike.

"Hey, I know you!"

"Hi, Jess!" I was excited she remembered me.

"So you made it after all?" I joked.

"What do you mean?"

"The trail, the PCT. I was wondering what had happened to you?"

"Oh, that plan . . . well that is a long story," she laughed.

Abby had a contagious laugh and genuine warmth. She was working as a research technician for a weed scientist and needed to rush out to the field. We exchanged numbers. It was really good to see her. We had crossed paths for only a short time in eastern Washington, yet it seemed serendipitous to run into each other again.

⟿

Oftentimes I would take walks with Fran in many of the local parks that decorate Corvallis; our favorite was the long slender park that gently curved along the Willamette River. Fran would dash in and out of mud pools along the river, and I would casually smile at the families and couples who passed me along the trail. I enjoyed being anonymous.

Jim and I stayed in touch, but it was hard. Regrettably, I mailed him the late-night letters you are supposed to write but not send when heartbroken. Our emotional conversations often disintegrated into long, drawn-out fights that weren't productive or helpful in any way. Neither of us really knew what we wanted or how to handle these complicated feelings.

Certain weeks I tried to be more social. Abby invited me over for dinner, which ended up being a Friday night routine. She had a new boyfriend, Steve, who was actually an old boyfriend from before graduate school. He doted on her and made us a fresh batch of cookies after dinner. They were living together, and based on their playful interactions, it seemed they had a long future ahead of them. I enjoyed being their third wheel. Their dog, a lively black mutt aptly named Muttley, became Fran's best buddy as well.

I attended various departmental gatherings and graduate student parties, but I found myself turning away, uninterested in the typical graduate school experiences. Jane had another graduate student, who was in and out trying to complete his master's degree, and a charismatic research assistant who was generous and always trying to get me out to do things. I wanted to embrace the newness of this place, but I was still haunted by betrayal. If I wasn't careful, my mind would wander to fictitious images of Jim and the other woman together. I recalled every time he told me he was "working late" or "too tired to hang out." Had our time together been a lie?

I was going crazy. Worst of all I couldn't trust my own judgement—how had I been so wrong?

⌁

After a few months in Oregon, an unsettled feeling crept under my skin like a persistent itch. I woke up crying most days and I didn't know who to talk to. Sometimes not for long, but other days I had to skip classes and stay home until the torrents ceased and the sobs stopped. I was centerless. As I approached my twenty-eighth birthday, an astrology-loving friend jokingly warned me that I was entering my Saturn return, a period of major life changes and disruption. I found that mildly amusing and, also, terrifying.

An old high school friend was studying to be a psychologist on the other side of the country and I reached out to her for advice. She suggested I see a therapist. No one in my family did that sort of thing. My parents had tried couples' therapy during their separation, but in both of their words it was "useless." Feeling like I needed to do something, I picked a woman's name randomly out of the phone book and made an appointment. She took my university insurance and was close to my house, so I could walk.

It was a sunny day, which made me realize how dark her office was by comparison. The receptionist was an older woman, and the waiting room looked like it'd been decorated thirty years ago. A classic rock station blared from a small radio above the receptionist's desk. I finished the necessary paperwork and waited, Led Zeppelin jamming in the background.

The therapist looked similar to the receptionist—were they sisters? Her office was slightly lighter since there was a skylight, although the skylight had not been cleaned for what was probably thirty years as well. The therapist was wearing a light blue suit, and I scanned the row of degrees carefully hung on the wall by her bookcase.

"Why are you here today?"

I fidgeted in my seat.

"This is a new process for me."

"Why are you here today?"

"Um . . . I'm not feeling like myself. I broke up with my boyfriend, well . . ."

I took a long pause.

"He cheated on me."

Saying it out loud made me feel something in my chest. I hadn't told anyone in Oregon what had happened, not even Abby, and I'd intentionally lost touch with most friends from afar, not wanting to tell them my newest news. Facebook was just being born at that time, and in retrospect I am grateful I didn't have a virtual stream of friends to update or the ability to debrief publicly. As a woman in graduate school, achieving more academically than my mother and grandmother yet constantly battling imposter syndrome, this humiliation hit hard.

I told her more about why I was in Corvallis and my goals for therapy. My owns parents' divorce was a subject I had been avoiding.

"But we are working it out and my dad cheated on my mom, but they are still friends and, well, they're divorced, but they worked it out and it's fine."

My new therapist stared at me intently.

"By working it out, do you mean you have forgiven your partner?"

"Yes," I said abruptly. I paused. "Well, maybe?" He wasn't really my partner anymore.

"Have you forgiven your father?" she asked, more intently.

My mind drifted to my cousin's wedding the year before in Chicago. A rather blunt uncle pulled me aside randomly as I was headed to the dance floor. "Hey, why can't you let your dad be happy?" I shrugged and walked away, half tipsy and also annoyed. I had maintained connection with my dad, but I admit it had been hard at times. From my perspective my dad was doing whatever he wanted without any regard for how it affected me, so I wasn't quite sure what my uncle had meant.

"Years ago," I snorted. "Things are fine. We are friends. He is coming to visit and see my new place. I told him that I just don't want to ever meet her. You know, the other woman."

"Does this other woman have a name?"

"I'd rather not say."

She smirked, and the session went on from there. Later she sent me home with a book called *Forgiveness Is a Choice*. I set it on the stand next to my bed. We had a few more visits before I decided I was too busy to continue.

When I went to live and work at the herb farm, my parents officially divorced after years of painful separation. Our family of four grew up in a spacious house where we could live alone together. In truth, we had all been growing apart for years.

It had been a shock to find out my father was having an affair after twenty-five years of marriage. He and my mom were both from large Catholic families where divorce was not an option. My parents had been together for so long, and while they had a close friendship, I could now look back and see obvious signs of faded affection. From the outside it was easy to rattle off why my dad had an affair (they had grown apart, they were not in love anymore), but as a daughter it wasn't easy to accept.

We were never clear on when the affair started. It was most likely when I was away at college or maybe in high school, but no one wanted to press too much to find out the specifics. He must have been around forty-seven at the time, which is statistically the lowest point in a human's happiness curve. Had he been unhappy?

One night over a winter break from college in Vermont, I was having trouble sleeping. Wandering downstairs for a glass of water, I heard a car running. I found my father in the garage, engine running. He was trying to kill himself, I thought. In the split second I had to think, I decided it was because he couldn't handle the pressures of work or fatherhood or something like that. Horrified, I slammed my hand against the window, yelling and expecting the worst.

Hunched over a cell phone plugged into the cigarette lighter, he quickly popped up and put his hand over the mic and mouthed "WORK CALL!" I stepped back, relieved and confused.

Looking back, I realize it wasn't a work call.

The secret came out during my senior year of college, and from then on, my father and I would go for periods of time, weeks or sometimes months, without speaking. I didn't know how to talk to him. My mother kept her daughters' interests at the center of their separation and eventual divorce. They always stayed civil with each other, and friendly, even though I knew my mother was deeply wounded.

I took pride in being my mother's steadfast adviser, a reversal of roles she resented. I had so many suggestions and answers for her. *It's time to move on, it's time to be your own person, embrace this new chapter.* I made it sound so easy, so Oprah-like, despite my own obvious lack of personal experience. My time as an intern and a graduate student bolstered my know-it-all qualities. Unlike my sister, who'd fled southward to LA, I was ready to fix my parents and make immediate sense of their divorce. I believed there had to be a sparkly, silver lining to all of it. I believed it had to make sense.

And there was the cautionary tale of Kelly. Another mother from our neighborhood, who had two daughters the same ages as my sister and me. Her husband had an affair and left her after decades of marriage and devotion. She was "fine" for several years; she even started dating again. Then one day she drove off a bridge. Betrayal can be brutal.

⟡

Perhaps that is why my own relationship failure was so piercing. Jim knew what had been going on in my family. I didn't hide it; in fact, I'd eagerly talked to him and sought his support.

Later, I would find out that the elders at his parents' church held several meetings and interventions to deal with him and our scandal. The details were fuzzy, and within his family it seemed like there wasn't much discussion about what had transpired. At least, no one reached out to me. I was gone, so it all could be swept neatly under the rug. I found it odd that a family who regularly and openly talked about spirituality was so averse to talking about betrayal.

I kept obsessing over the facts: I knew her. I knew her husband and son. They were friends of his family. I had played Legos with her kid.

In biblical terms, the fall from grace or the fall of man in Genesis explains the transition of the first man and woman from a state of innocent obedience to God to a state of guilty disobedience. When we falter in life, we automatically think that we are at fault, that there is something wrong with us, or that there is someone to blame, especially ourselves. However, Joseph Campbell says, "When you stumble and fall, there you will find gold."

I didn't find any gold, initially, only the worn beige carpet in my rental house, the new bead on my string of temporary homes.

✦

Although I had a small field site at the research farm, it did not really feed my urge to be working on the land. Like most land-grant research farms, including the WSU farm, the OSU farm was conventionally run, tidy (since all the weeds were chemically managed) and well-organized (neon flags marked the various research plots). The OSU farm grew a mixture of fruit trees, filberts, vegetables, and turf grass. It had at one point been a working farm, so there was an old farmhouse and various metal outbuildings that had been put to use as equipment storage and research space for observing specimens, sorting through soil, counting seed, and drying and weighing plant material. The original barn had fallen sideways years before. Activity on the farm ran at a manageable if sometimes sluggish pace, since no one was in a hurry to harvest for market.

The farm manager, Roger, held a lot of animosity toward Jane. Abby warned me about this since she was at the farm often helping her boss with his many field trials. Roger and Jane were polar opposites personality-wise, although they shared an argumentative streak. On the days she did make it out to the farm (he commented on her long stretches of absence since getting tenure), Roger would stand up and clap to celebrate her appearance, an obvious jab. She was never amused, and would regularly roll her eyes and avoid talking with him at all. I imagine he was never so bold with her male colleagues.

While I was enjoying running my own schedule, the field season passed, leaving space for many questions about my research project's direction. I scoured the campus for Jane, and she was harder to find than I had expected. My previous adviser had tracked my every move, but this new mentor-student relationship felt different. I knew she had jumped through a lot of hoops to gain respect (and tenure) in the scientific community with her longtime and often outspoken advocacy for organic agriculture. I admired her greatly, although I sometimes cowered under her boldness.

Many of the research technicians in the department snickered when I asked them if they had seen her in the lab or around the department offices. It was common knowledge, apparently, that newly tenured professors were often hard to find. After five or six years of researching, writing grants, graduating students, and proving their worthiness to the institution, many would use their official acceptance as a chance to retreat from all the pressure.

Carl was usually in his office and acted as a sounding board for my ideas about my project and research questions. I used his lab for most of my sample analysis, since Jane did not have a technician, and, as a newer professor, her program had considerably less funding. They had a strange friendship that was borderline flirtatious. However, Carl was married with two kids and seemed quite conservative, while Jane was by all accounts eccentric. I found that the balance of the two of them was helpful to me, and I appreciated how friendly and relatable they were in their own unique ways. They acted more like academic midwives than the dictator model I had grown accustomed to in my master's program.

A few times Jane invited me off campus to visit her farmer friends. Once we stayed overnight to help a couple with their squash harvest. Being outside, doing physical work again on a beautiful little farm in a remote valley, was invigorating. Another time we met up with a whole gaggle of farmers at a nearby hot springs. They had formed an unofficial conference that was farmer-led. Everyone met and decided what they wanted to talk about, instead of following a plan set by university officials. Then groups broke off into different clothing-optional soaking tubs to discuss their ideas and questions. I distinctly remember having a very insightful, yet uncomfortable, talk about the benefits of no-till agriculture while I was bare-ass naked.

I did find comfort in the times when I could meet with Carl and Jane together. We would sit in Carl's office and often have very hilarious conversations about plant diseases, their own experiences as graduate students, or my stories about farmers in the Skagit Valley. I always wondered if they could tell I was homesick.

✦

At a party of graduate students and some faculty, I talked at length with an instructor in the crop and soil science department. He had had a previous career as an internationally famous rock musician before falling in love with agriculture and eventually returning to graduate school. In my drunken state, I proceeded to go into embarrassingly excessive detail about my previous relationship.

The man patiently listened and then smiled. "You should be thankful that you are through with that 'let's farm together' phase before you lost any real money."

He looked at me deadpan. He wasn't joking.

"What do you mean phase?" I asked.

"You are not the only one who has had that dream, and let me tell you, that is all it is, a dream. It never works emotionally or financially, and somebody usually ends up getting hurt."

I stared at him blankly and then went to get another drink. Part of me felt unnecessarily cautioned and part of me felt that he was right. I had been projecting my dreams onto my relationship with Jim, and in reality I had no idea how we would work together as farmers, let alone business partners. I wasn't planning on playing the stereotypical role of a farm wife, cooking and cleaning and child-rearing while he did all the actual farm work. I'd never really known what he wanted.

✧

One hazy summer afternoon, while I was bringing soil samples from the research farm to the lab, I noticed Carl and Jane driving off together from the agriculture building parking lot in her dusty, dented truck. In the previous week, I had tried to reach both of them about my project, but my emails went unanswered.

I walked to Carl's lab down a hallway of white crackling walls and vacant offices. State budget cuts had reduced the number of faculty in the sciences, especially in applied science, like agriculture. Some research programs, such as those studying grains and potatoes, had a lot of external support from corporations and crop commissions. But overall, despite the growing public interest in food systems, the facilities screamed underfunded. Except of course for the new engineering computer science buildings—they were new and shiny.

Along Carl's hallway in particular there were only two active offices and labs, and the rest were empty or filled by students or technicians. Old glassware, chemicals, and other supplies had stacked up over the years, and each space felt cluttered and weathered. You could almost imagine the heyday here, men in clean, slick haircuts and pressed white lab coats, making discoveries. Now there was a dullness to the wall color that made the fluorescent lights all the more jarring.

Lila, Carl's research assistant, was in his lab. I told her I had seen Carl and Jane leaving together and she rolled her eyes. Lila had short hair and walked with a deliberate step, like a Buddhist monk. She had worked in various labs for many years and could do most anything. Unlike a lot of the tenured professors, she liked doing the actual work of research, and she was good at it. We had become friends as I came to her with all my small questions about my projects, the details Carl did not want to deal with or have the answers to. She handled all of the daily operations of his lab's research at that point, and she had a keen curiosity about all organisms, especially fungi. She loved microscopes and her technical help and advice was invaluable to me.

Carl was established enough that his main work focused on writing grants and papers and teaching an undergraduate course. Although she never went on for an advanced degree, Lila had done Carl's work for so long she could have earned an honorary PhD for her research.

While she wasn't one to gossip, I could tell something felt peculiarly odd, and Lila was noticeably irritated. She paced back and forth in front of me.

"There is something going on with Carl and Jane," she said bluntly.

"What?" I murmured.

"You might want to look into another project. I think they are a thing, and this could get messy, and you shouldn't be put in that position as a student. And, well, I don't know what else to say. You shouldn't have to deal with this."

My cheeks started to turn red and my arms felt itchy. Maybe I had picked up on the clues? They were rather flirtatious in meetings. Jane and Carl's personal lives were not my business, but what if they broke up? What if they stopped speaking? Then my project would be in limbo. Wait—an affair is immoral. Was that what I believed, and why it bothered me? I felt my shoulders turn to stone, and my neck slid into my back as if someone was about to hit me over the head with a baseball bat. I had been walking around that way for months.

❦

I met with Jane at a corner coffee shop in town. The old standard that had seen piles of books and computers, first dates, and nervous freshman sit at its wooden tables for decades. I wanted to clear the air. Why had she been so hard to reach? Was this project going to work? After we went over a few details I just blurted it out.

"Is there something going on with you and Carl?"

She was shocked. Her eyes began to look down at me from over the top of her glasses—which was never a good sign.

She let out a sigh and with a heavy breath said, "No!"

"I just needed to ask, because that would be a conflict of interest, I think. For my project."

She wore a subtle yet coy grin, almost like she was going to laugh. Or maybe she was insulted that I'd asked her about her personal life. She changed the topic quickly, and I was left to decide whether or not I should believe her.

❦

I didn't know what to do. Could I keep doing my work here with the echoes of Jim's betrayal and my parent's marriage on my mind? The act of doing science offered no balm to my aching insides. Who could I trust?

In my rental, with Fran by my side, I read everything I could get my hands on to feel better. Poetry, memoir, self-help books by the dozen. I finally flipped through the book the therapist had recommended. While loosely focused on my laboratory and fieldwork, I found myself wandering into the back row of lectures on religion and spirituality. Speakers like Marcus Borg, a progressive Christian scholar, and Kathleen Dean Moore were particularly insightful, especially Moore's discussion of an ecological ethic of care.

Karen Armstrong's spiritual memoir, *The Spiral Staircase,* was helpful too. I realized that Armstrong, to my mind a mogul of religious publishing, had once been a nun as well as a PhD dropout (a result of highly sexist conditions) before finding her calling as a writer. She writes, "We should probably all pause to confront our past from time to time, because it changes its meaning as our circumstances alter."

After confronting Jane, a strange malaise came over me that I tried desperately to shake. I needed to make meaning out of all this betrayal, the rafts of isolation. Technically, I was Catholic, but I didn't have a strong spiritual tradition to fall back on or limit me. I was trying to gather rather than shed. It would take me years to put my finger on it, call it what it was: grief.

True, no one had physically died in my life yet (thankfully), but I was traumatized by loss nonetheless. Jim's sister-in-law, who had become a friend, had lost her entire nuclear family. Her sister died of cancer in high school, and her parents passed away to long-term illnesses, one right after the other, when she was in her mid-twenties. Her losses felt unimaginable and I marveled at the grace with which she managed to carry on. How could our experiences possibly have the same name?

However, the dissolution of a family bond means something too, and I had been unknowingly avoiding that pain by escaping from it in farm fields and old barns, trying to create a new story for myself that wasn't so hard to accept. An adult child grieving the divorce of her parents may not be as hard as grieving a death. There is, after all, still time to create new memories and rituals; however, there is always the dark and painful reality that things will never be the same. Divorce splits the very ground on which the children (regardless of age) stand; to try to find footing after that kind of break often becomes an unconscious, and sometimes lifelong, search. In my case, I launched into my adult life in a frenzied state, seeking and exploring, moving all the time, unaware that what I really needed to do was heal.

⟿

Gathering Together Farm, outside of Corvallis in the small town of Philomath, is made up of a patchwork of parcels quilting the banks of

Marys River. Many of its fields are lined with ancient willows and oak trees, and oftentimes the fields flood and disappear for a good length of time.

It is one of the oldest organic farms in Oregon, and a founding farm in the distribution of organic produce in the Pacific Northwest—an industry that has grown exponentially in the last twenty years. They grow a wide range of crops, from tomatillos to carrots to winter squash, and they curate the most ornate and developed farmer's market displays I have ever seen. They take their produce as far as Portland and Eugene and also service many families in the Corvallis area with their weekly CSA.

Gathering Together has a farm kitchen with coffee and homemade potato doughnuts and a lot of other delicious culinary offerings. John, the owner, had been a chef before he started farming with his wife, Sally. The farm began when the idea of sustainable, small-scale agriculture was something only hippies knew about or attempted. During the fall I arrived, Jane had a few research plots on the farm, so I got to spend time there. As spring warmed into summer, I kept thinking of ways I could be at that farm more often, away from my research plots and empty house.

On Monday afternoons I volunteered to help pack CSA boxes in return for free produce. It brought me to the farm every week for a few hours, a welcome change from my graduate school schedule. I'd meet up with the CSA crew on the covered landing behind the farm's restaurant. We could vaguely see the river meandering behind us through the trees. We stayed cool and out of the late summer sun, standing on the shaded pavement surrounded by crates of produce and lulled by the calming hum of the cooler.

The crew consisted of a few farm employees and a few work-trade people like me. One woman was the wife of a chef who'd just started working at Gathering Together's on-farm restaurant. She wanted to get involved in the place where her husband was getting his culinary foot in the door. They had just relocated here from a big city and a fancy restaurant. The farm-to-table restaurant concept was becoming a draw for even the most high-end chefs.

Packing the boxes was challenging for me, since all of the produce was so gorgeous. I just wanted to look at it and admire the colors, especially the rich, red heirloom tomatoes. We would soon get into a rhythm packing boxes, and the company of other packers made the time go by quickly. We were each stationed by a different vegetable, and as each large blue tub slid down the table between us, we added our allotment. Heavier produce was packed at the bottom, and the more delicate items, like the prized tomatoes, were placed carefully on top.

Sally, thin with short brown hair and a friendly smile, ran the packing shed and drifted in and out, keeping an eye on everything. You could feel her peeking over your shoulder, making sure you weren't giving too much of one thing or putting it in the wrong place. Presentation mattered. If we were short-handed, she would step in and help out. She seemed to never stop working—or moving, for that matter, usually wearing a farm logo T-shirt and a pair of dusty jeans. Sally was three times faster than anyone there and could pack with her eyes closed.

On certain days I was able to take a box or two of tomato seconds home; I brought home the blemished fruits in gluttonous amounts for a one-woman household. Entranced by the red glow coming out of the box in the corner of my kitchen, I searched cookbooks for ideas. I didn't want to waste them. Sauce, for sure, maybe some homemade ketchup, too, if I had time. I'd been around the process of canning in my grandmother's kitchen, but that would be something I had to relearn. I was out of practice.

Consulting *Better Homes and Gardens New Cookbook*, the one my grandmother had given me, I armed myself with all the tools, from the tongs to the mason jars, that the hardware store had on display. Apparently, I wasn't the only one in this town with a penchant for storing food. Supplies were bountiful and affordable, so I had no excuses.

I chopped the tomatoes and cooked them down. Then I sautéed onions and garlic and herbs and added it to the mix. The smell was rich, and I couldn't stop tasting spoonfuls. As the sauce simmered, I filled the big black canning pot with water and waited for it to boil. After the sauce was transferred to the washed and rinsed mason jars, I carefully placed the caps and screwed on the lids. Wiping the edges clean of any spilled sauce, I placed the jars into the boiling water using my now trusty tongs. Those things did come in handy.

After their allotted time in the water bath, I pulled each jar out and placed it in the orderly row on a folded kitchen towel. They were majestically red and steaming, and they looked like little soldiers ready for duty. I waited patiently to test the lids, and the occasional "pop" was evidence that there had been a good seal and enough pressure. The lids were officially on!

From then on, instead of scarfing down a bowl of pasta while standing in the kitchen, I started to sit with myself. I couldn't eat the sauce that had taken me hours to create and store as quickly as the cheap store-bought jars I was used to buying. I couldn't forget the hours it took John and Sally to start that seed, tend it in the greenhouse, plant it in the ground, water it all season long, harvest the fruit, and get it washed and packed and ready for the CSA boxes.

Cooking and eating alone began to be my medicine. Breathe, chop, sauté, mix, knead, bake. All those motions became rituals. I could get lost (and away from my monkey mind's chatter) in the processes of canning food and making things, like bread, from scratch. In college, I had started cooking for myself in small doses, the *Moosewood Cookbook* as my guide, and I appreciated the time and space to focus on food. I lusted over Gathering Together's bounty, which was overwhelming at late harvest. Squash, kale, potatoes, carrots, kohlrabi. If there is such a thing as vegetable porn, it is a Gathering Together farm stand in mid-September.

Pema Chodron writes, "So even if the hot loneliness is there, and for one-point-six seconds we sit with that restlessness when yesterday we couldn't sit for even one, that's the journey of the warrior." I didn't consider myself a warrior, but I had somehow arrived at a more settled place in my life. A pause. I was not planning to be in Corvallis forever. As I started spending more time with myself, a new sensation came over me: something indescribable, yet familiar. I was re-meeting myself, again, but for the first time. And I liked this new friend.

❧

During my yearly visits to Wisconsin, I'd watched my grandmother love her home and family, and I wondered if I would ever find the same comfort in my life. She loved and accepted the life she had, even though most of it was a tapestry of someone else's choices. She had been expected to get married, support her new husband in the navy, and then produce many children after the war; however, she never seemed resentful. She embraced her life, her church, her part-time job as a librarian. My grandmother mastered her kitchen and still had time to care for her own dying mother and keep up correspondences with all of her first and second cousins. They say that contentment is falling in love with your life. She was, by most standards, content.

I thought about my grandmother's decision to live alone after her husband, Papa, passed way. Initially, we were not sure she would want to be alone with all those memories of her life with Papa. My mother's siblings had all migrated to the western reaches of the United States, and they were all secretly hoping their mother might come out this way too. But Grandma wasn't going anywhere. After her grief lessened, I think she learned to love the space and her own quiet company.

When I visited, I observed her routines. She would wake up in the morning, watch the birds, do tai chi exercises and her crossword puzzle, and the day would begin. My grandmother had a resident opossum that lived under the back deck. For years, the furry creature made a routine of coming up to the back door in evening sunset light, standing on his hind legs, and peering in through the glass. She never offered him food, but he checked on her anyway. They had some sort of bond none of us could really understand. In some ways, the family was grateful it was there to check on her.

In the evenings, summer or winter, we would walk down her hill, past the pond edged in red sumac. Her neighbors, almost like family after all those years, would wave as we passed. She was somewhat of a celebrity on her street, known for her tasty baked goods, generous spirit, and good humor. Some days the mosquitoes would be unbearable and we would turn back. But if we timed it right, we could miss their swarms coming off the water.

In winter, when the trees were bare, we could see Lake Winnebago and the light reflecting off its shimmering water. In the summer, the wild *Echinacea* and *Rudbeckia* species would bloom along the restored native meadow at the park, and my grandmother and I would stroll among the pink and yellow blooms.

I spent most of my life going to that house, and the sameness of it was important to me. While my family moved, and as I grew through various stages of awkwardness, it felt like that house stayed steady: the long arching walkway, the fireplace and expansive bay window in the living room, the dining room and piano, the basement freezer and dress-up trunk, the musty-cozy smell of the downstairs twin bed, the bird feeders, the cabinet of photo albums. We would measure our height by taking a picture alongside the statue of Chief Redbird at the nearby state park. A park, I later learned, created around the sacred effigy mounds used by the Winnebago (or Ho-Chunk) tribe for ceremonies and burials. Our family history was layered on top of the long and deep history of the land itself. The limestone cliffs, permanent but slowly eroding, were markers of a scoured memory.

My grandmother held me together in many ways, and I admired her life. I loved visiting, cooking, playing cards, watching the news, going to the Piggly Wiggly. Her domestic life felt almost like what I imagined enlightenment to be. When we sat at her kitchen table, we simply enjoyed each other's company. I didn't have to be anything or anyone. I could just be. For me, that was groundbreaking.

I like to think about my grandmother as part of the same Silent Generation as Wendell Berry. She didn't have a lot of choice, yet she never seemed to carry much regret. She raised five kids, she buried her husband, and she continued to live alone for another ten years. I imagine she wasn't happy all the time, but from my point of view she leaned into her life and her home. Like Wendell Berry says, she was "in place and free."

In response to an essay in *Harper's* about not using a typewriter, Berry received a rash of letters criticizing him and accusing him of exploiting his wife, who was his long-time caretaker, editor, and typist. In his essay "Feminism, The Body, and The Machine" he defended himself, saying, "I do not believe that there is anything better to do than to make one's marriage and household, whether one is a man or a woman." While I'd long coveted a certain domesticity, I'd never quite stopped to process how that might play out in daily life. When I thought that Jim and I would start a farm, I never really thought about marriage at all. I had always imagined having a partner on a farm, but I was starting to question that dream. Perhaps one of the reasons I thought I loved Jim was because he was already living the life I wanted for myself—in the place I loved.

I didn't want to marry the farmer; I wanted to be one.

As a young woman pursuing an advanced degree, I felt compelled to live out all the choices my grandmother did not have. Yet at the same time I craved domesticity and admired the life and satisfaction that came largely from her place-making, her behind-the-scenes work as a wife and mother. This was work I knew nothing about. On the one hand, I had opportunities ahead of me as a nomadic professional; on the other, I was spending the majority of my time daydreaming about farming in the Skagit. I loved the idea of doing simple, elemental things—baking bread and planting seeds. Was I looking to give up my progress as a woman to live out some antiquated fantasy?

There was something about how my grandmother lived, how she made things and took care of things, that predated the consumerism I knew and practiced, the overconsumption my generation was pushing back against. It was also interesting that my own mother had no interest in the skills my grandmother, and even great-grandmother, had lived by. Knitting, baking, sewing, weaving. Cooking especially seemed oppressive to her. She showed care in her own unique way, and I absorbed the fiercely independent way she managed our household and finances. However, I still yearned to reconnect with this lost lineage of matriarchal knowledge.

As the summer rolled along, I was not only in limbo about working with Jane and Carl, I didn't know whether I wanted to do research at all. I missed farming. I knew I had been raised with a lot of privilege and I didn't want to squander it by leaving this path, giving up the educational opportunities my grandmother never had.

⚘

In crafting my dramatic letters to Jim, I had unknowingly jump-started a writing practice. I even signed up for an online poetry class behind Jane's back (what a renegade!). It was a necessary shift away from my struggling-to-get-off-the-ground bean mold research project—and all of the science writing I had been trained to do. Poetry, I found, was the best vessel for connecting with myself, with this longing I felt. That earthy thrill that bored under my skin at the herb farm and later at Gerry's homestead.

I had never quit anything before. I had never left anything unfinished. Now I was trusting some deeply rooted flight instinct, like the migratory swans that return to the valley every fall. I decided to leave my PhD program, a year and half after moving to Corvallis. I knew I couldn't stay. I was grateful for my friendship with Abby and would miss being close to Dee, but something wasn't right.

I was listening to the voice buried deep inside me like a small, shiny seed.

Sheep

SHEEP CAPTIVATE ME. I enjoy watching the expressions on their soft, guiltless faces. Although many people think goats are smarter, I believe that sheep have a lot of emotional intelligence. I never quite know what they are thinking, but they always seem to be aware of what I am doing, as well as the activity of the other animals on the farm. Also, they look out for each other. If something is wrong with an individual ewe, unrest permeates the flock.

When I was a freshman in college, I took a nature-writing class. The professor, who also moonlighted as a Vermont shepherd, asked each student to spend a night observing and assisting spring lambing on his farm. He had a small bunkhouse above the barn floor, and when it was my turn, I spent the night with two other students nervously pacing behind suspect ewes. He was a very trusting man.

Both excited and nervous, I was ready to pounce on the first ewe that went into labor. She pushed and pushed, and through the gooey film expanding from her backside, we saw a small white face protruding. Then, surprisingly, she stopped pushing, and the peaceful, silent face just seemed to float there in perfect balance between two worlds for what felt like hours. Was it alive or dead? Had it passed in utero? I shrieked and danced with joy when the full body finally emerged (no thanks to us) and began moving—a new life born to this strange and captivating planet. I didn't realize it then, but that experience would serve as the start of my fascination (and obsession) with these loyal, caring animals, as well as the ritual of spring lambing.

During college I also worked at a small, rural inn. The owner tended a flock of wool sheep, and annually she would have their fleeces made into beautiful yarn and blankets. I fell in love with those sheep and eagerly volunteered to feed them on the days I worked. One of the lambs at the inn was named after me. That adorable little white lamb grew up to be quite a force—a robust and fierce creature who stayed at the inn far beyond the average sheep's life expectancy.

Now, over fifteen years later, I find myself out in the barn feeding my own flock, East Friesian and Lacaune with a little Romney and Dorset mixed in somewhere along the way. As the lambing seasons continue, we encounter complications and even death. As Berry describes in one of his Sabbath poems:

> Near winter's end, your flock
> Will bear their lambs, and you
> Must be alert, out late
> And early at the barn,
> To guard against the grief
> You cannot help but feel
> When any young thing made
> For life falters at birth
> And dies.

Yet, the miracle of witnessing each lamb drop is addictive, a rush of adrenaline and spiritual insight all in one precious moment. Sometimes it is too much for my heart to handle.

The sheep had been an accessory on the farm in those first years. The house and barn were part of a forty-acre cow dairy that started in the 1930s, but nothing had been milked here since the 1970s. I experimented with a homestead bucket milker, and although the sheep were reluctant to enter the makeshift wooden stanchion that holds their bodies in place while milking, they eventually got the hang of the routine. In our home kitchen, I figured out how to make some flavorful fresh cheese and yogurt, and I started daydreaming about a creamery. The sheep herd tripled in size. I began planning for a new and bigger sheep barn that could shelter the whole flock.

Sometimes I stop and wonder if it is a crazy dream to think about milking and making cheese commercially. Most of the dairy farmers I know are doing the work because their parents and grandparents did. In Skagit County, the number of larger cow dairies decreased from thirty-seven in 2006 to twenty-nine in 2014, a pattern of decline that has become the norm thanks to rising land prices, the cost of raising animals, and our nationwide propensity for cheap food.

However, small and micro-dairies with anywhere from ten- to two-hundred-head herds are growing in popularity, especially in milk-rich states like Vermont, Wisconsin, and California. Many of those small dairies make their own cheese as well as milk-based products like butter and yogurt that can be sold directly to customers at higher prices. All of these value-added products, farm goods made with raw material from the farm that bring direct income to the farms, can be a solution to the instability of milk prices. There are on average 450 artisan cheesemakers in the United States, and that number seems to rise each year; small farmstead dairy farms are producing amazingly delicious, diverse, and unique products.

Cow and goat milk cheeses are the most popular and common of the artisan milk-based products, although an interest in and appreciation for sheep cheese is growing. I have always loved European sheep cheese like Manchego and pecorino, and our local market carries a delicious Israeli sheep feta. Domestic versions of these cheeses are harder to find since there are only a handful of sheep operations around the country.

I want my sheep to be working animals. Raising sheep is not cheap when you add up grain and hay costs as well as shearing expenses. The sheep already give back in a myriad of ways: companionship, manure for the soil, meat, and wool sold within our community. The possibility of making sheep cheese is alluring to me, and potentially lucrative. Cheesemaking is work that we could do ourselves without the help of cheap labor. It is a farm product we could be proud of, something we could make with our hands.

I know the work will be hard, but the rewards—good food and happy animals living in balance with the land—might be what matters most.

Skagit Valley, 2008

WHEN I RETURNED to the Skagit Valley in early spring after a year and a half away in Oregon, it didn't take long to settle into the familiarity of houses and roads, the sameness of scenery. This was the first time in my life that I had chosen to come back to a place. As I child I'd had no choice but to let go of our familial roots in Wisconsin while my father chased his career. As a young adult, my only direction had been forward—to college, jobs, new experiences. But now something was calling me, like the scent of a riverbank to the salmon that return to the same river every year.

⊷

In the fragile weeks after leaving Corvallis, I surprisingly ended up at my dad's house outside of Seattle. I had broken the lease on my rental in Oregon, put my things in storage, and needed to land somewhere. My mom was not interested in having a dog at her house and I knew my dad would be traveling a lot, so even though things were sometimes tense between us, I'd have my space.

We shared some nice dinners together, and I took walks along Lake Washington with Fran and did yoga at a studio down the road. I enjoyed the slow pace, the break in time. It was both comforting and odd to be around my dad, in this strange new environment. Although he had bought this modern two-bedroom home mostly furnished, it was interesting to see which paintings, like the large Hugh Cabot cowboy, and décor he had acquired in the divorce. He was settled in, but there were piles of work papers in kitchen drawers and expired cartons of milk and stacks of to-go containers in the fridge.

In many ways I felt like a failure, walking away for school and a chanceat a new life in Oregon, but there was a twinge of hope inside me too. Love comes in many forms, and I realized I was longing for a place more than a person.

⊷

I found a small apartment above a barn and after a month with my dad moved northward, back to the valley. It felt like a sign, since rentals were hard to come by in Skagit, especially one this private and new. My landlords were a retired WSU extension agent, Erv, and his gregarious wife, Julia. Extension agents work for both the university and the county to provide trainings and information to farmers, gardeners, and landowners. Erv had a long career with WSU and was proud of the work he had done. He himself could have been a farmer, but he was able to share a lot of his knowledge by taking this path as a public servant. Erv and Julia were now grandparents, very religious and active in their church. I was a quiet tenant with no partner or semblance of a nightlife, so we got along just fine.

Over the spring and summer, I had taken on more research work at the newly renovated WSU center, where I had done my master's degree, which was just down the road from Erv's apartment. In the time I was in Oregon they had created an entirely new facility with state-of-the-art laboratories and offices. While many of the other land-grant research centers around the country were losing funding, falling apart structurally, or closing down, this one was being saved, which spoke to the support for agriculture in general in the Skagit Valley. I was happy to be there again and to be part of the final stages of revitalization.

The early spring that I arrived, Erv was working on weaving daffodils into a huge cross, almost taller than he was. It was part of their church's "flowering the cross" ceremony, which symbolized darkness, pain, and hopelessness transformed into something beautiful, new, and alive—a celebration of Easter and the resurrection. The amount of yellow was almost blinding, but I admired the beauty and the care with which he attached each delicate bloom. In the fields around us, daffodils and soon tulips were blooming in multitudes.

While staying with them, I started to look for available land. I didn't know how long I would be around, but we had a good relationship and I appreciated the enthusiasm with which they cared for their yard. Erv was always up at 6:30 a.m., either weeding or tending to other chores. After 9:00 a.m., he would pull out the big equipment, like the mower or weed whacker, to spare the neighbors and me, I guess, although I was up early too. His tenacity was admirable. Outlined with flowers and tidy rows of vegetables, his land was pristine. In the mornings I'd cradle my mug of coffee with Fran on my lap and peer out the barn windows at his masterpiece—and also toward the farmland beyond it.

Over the years I had stayed friends with Henry, the farmer Jim worked for, enjoying long talks with him about farming and land. Whether or not Henry wanted it, I became his devoted mentee. I think he was shocked when he found out what had transpired in my relationship with Jim, but Henry had enough forgiveness for a thousand sons. Through all of my relationship back-and-forth turmoil, Henry was patient—and kind enough to stay in touch with me. He knew farmland in this county did not come up for sale often. Henry kept watch.

We met up at Fortune Restaurant in town because he loved Chinese and Japanese food, and he also wanted to keep our meetings secret from Jim, I think. Before I returned, Jim and I were on good terms, relatively, but once I physically arrived, things turned icy cold. My feelings toward him were confusing, hate pushing up against care, but it was increasingly clear that his attentions had wandered. This was, eventually, a relief. I noticed an odd thread of comments on his band's MySpace page from various women— we weren't together, but I knew there was still a lot I didn't know about him. The wondering—about where he was or who he was with—was toxic. It still wasn't clear when (or if) he ever told me the whole truth. There didn't seem like a way to be friends, and I couldn't let myself slip back into the stew of self-doubt and paranoia I experienced in Oregon. I needed to cut off completely.

Henry was wearing a clean, white T-shirt, jeans, and a blue baseball cap that was shifted slightly to the side. He had long, nicely trimmed sideburns, and wisps of his salt-and-pepper hair escaped his hat. Henry's father had been a farmer both here and in their native Netherlands. Henry had wanted to do other things with his life, like woodworking, but the farm called him back. Like a loyal son, he took it on, and the farm flourished.

He always told me people weren't useful until they were thirty-two. Coincidentally, that was the age he was when he returned to take over his dad's land.

"What are you looking for, Jessica?" he asked me.

That question always gave me pause.

"I think I want five to ten acres, and I want to start growing vegetables and some herbs. I like fennel bulbs and kohlrabi and radicchio, I think there could be a lot of interest in these things."

He agreed with me, adeptly maneuvering his chopsticks.

"Make sure you don't start too big, and you don't want to be too close to the river."

His comments were always swift and to the point. I wasn't quite sure why he was helping me, but I was grateful.

↜

My dad called me one fall afternoon as he was driving back to Seattle with a few of the things he had left in Idaho over his decade of extended visits with a woman I had never met. I was convinced it wouldn't last. Although my mom never put any mandates on me, I wanted to protect her. As the daughter, not meeting the other woman was the one thing I could control.

I was in my apartment that day with Fran, looking at real estate ads, when he called. As we talked it occurred to me that my dad and I were both going it alone now, but together, and for the first time in a long time that was strangely reassuring. We had been talking more over the past year, especially since I had stayed with him. He was always surprisingly supportive of my interest in farming, a world far apart from his life in marketing. He always managed to visit me at my eclectic abodes too, from the intern house in Williams to my purple house in Corvallis, albeit some visits were more strained than others. Despite his indiscretions and the rift they created between us, I had to admit that he always encouraged me to follow my passion.

And he was loosely aware of all that had transpired in my love life too.

"Jim was a good guy, I liked him," my dad stuttered into the phone.

I sighed.

"There will be others," he mumbled.

I didn't respond, and after a long pause we moved on to talking about the Mariners, their disappointing season or something like that. My dad always assumes I know more about professional sports than I actually do.

When he got back to Seattle, we spent a day together looking at new apartments for him. He was selling his house and wanted to start all over. It was a sunny Saturday in October, and the city was on fire with autumn leaves. As a passenger, a sidekick, I grew calm and nostalgic. I had never lived in Seattle, but I had spent so much time there growing up. It was in many ways an extension of my home—my city, if ever I were to live in a city. I was light, joyful, curious.

The breeze spurred whitecaps across the surface of Lake Washington as we soared across 520. Boats bobbled in the wind around the floating bridge that connected my childhood home to the east and the city that I loved. We were both starting over in our own strange ways, and we shared a comforting silence in the car. I did not know very much about the life he had created with the woman he had left my mother for and now that it was over, there was no reason to ask.

We visited an apartment at the top of Queen Anne, a townhouse a few blocks off of Broadway, and then a tall condo in the hills above Alki Beach in West Seattle. I wasn't sure where my dad would land, but I could feel he was ready for a brand-new start. We both were, and I was glad for the chance to get to know him again in a new way.

That evening as I drove back home to the Skagit, a wave a peace came over me. The sun was setting as I descended down I-5 into La Conner and Mount Vernon, past Starbird Road, into the valley. It was harvest time, and in the fading light I could still make out cornfields and rows of potatoes. It was a beautiful sight, and I relished the thought that so much good is grown here, continually, year after year. For some the work is not a choice; it is what their family has always done. For me it was an aspiration, a budding chance to learn and grow and hopefully make some goodness out of the rest of my life. I sent my intentions into an orange-tinted sky fading into the brown fall fields.

⟁

My search for land was tiresome. As Henry had warned me, most farms in this agricultural valley stay within families. Word of mouth is the best tool. As I settled into my search, I started scouring real estate websites and contacting realtors. Henry continued to pass along inside scoops that were helpful, but didn't lead to something I could afford. My mother (with her impeccable credit) was willing to help me. She had always managed our family life and was more than prepared to be financially independent from my father after the divorce was final. However, land prices were high.

For a time I was in conversation with a dairy farmer who was eager to retire from milking and spend more time on his fishing boat. We worked out a contract that I would be able to afford only with a massively large loan. We met several times, and I loved the location of the farm, but I ultimately panicked about the amount of debt I would be taking on (during a huge recession), as well as the oversized and aged dairy buildings that I was going to have to figure out how clean up and repurpose for my scale of operation. As someone who'd never owned a house, the enormity of that dairy farm and all its buildings was daunting.

The next place that fit my price range and criteria (livable house, five-to-ten acres, good water) was a ten-acre piece down the road from Henry's farm. It had a one-story rambler house with purple shag carpeting. The land was long and sandy and made a perfect rectangle. It was far enough away from the river to be safe from flooding, but close to town so I wouldn't feel too isolated. The house was not too far from being a total teardown, so I made an offer. My hand was shaking with every page I signed in the agent's small office in town.

The farmer who had owned the farm and lived in the house in his retirement had recently passed away. His son, eager to make some inheritance, immediately (within a day) turned down my offer and would not budge. (Several years later, that property would sell for less than I offered.) For many children of farmers who do not themselves farm, selling land at a high value is the only way they can get any sort of money or inheritance from their family. In this valley, farmland used to be worth $1,000 per acre; now it's more like $10,000. The draw to sell the land to the highest bidder is strong, which makes farmland preservation and the availability of cheap farmland for young farmers very complicated.

As it turned out, I hunted for places throughout the valley for over two years, patiently waiting and looking, contacting strangers, and asking as many questions as I could. With all of the uncertainty following the economic crash in fall 2008, waiting seemed like the right thing to do. Jim's sister-in-law owned a house in town and was in the process of moving away. Eager to have a yard for Fran (Erv wasn't too keen on a dog galivanting around his beautiful gardens) and a bit more space, I moved into her house on the hill in town and found a roommate—another single woman, also recovering from a recent breakup. Maggie was an art teacher at a local private school and tirelessly worked long hours. She was kind, quiet, and devoted to Jesus, and exactly what I needed.

⊸

Back at the research center, I worked as a research assistant and was able to salvage the PhD program I'd begun in Oregon. Remembering Gerry's advice, "Get as much education as you can," I thought I might want to teach someday or do research on my own piece of ground. If I didn't have the land for a farm now, I would have it eventually. In the meantime, I could get paid to finish my degree, and I wanted to complete the academic path I had started.

This time I found a project on raspberry diseases and soil health, and the research center was comforting in many ways. I enjoyed being back. Even though the entire campus had been completely renovated, the people and their routines were familiar. The tree house, used for small meetings, was still there as well as the tired metal Quonset hut. There was a family-like atmosphere, and several of the longtime employees, the ones born and raised in the area versus the newer faculty who had moved here for tenure-track positions, looked out for me—sharing plums from their yard or giving me a ride when my car was in the shop.

Henry let me use a small patch of ground near his house. I was technically renting it, but he never asked for money. Instead, I grew an odd mix of things he didn't already grow and tried to harvest and sell it to local restaurants in my free time. Most days it was a wild mess. He would secretly send his crew out to help me with weeding, knowing I was in slightly over my head.

For my research project, I'd been hired to examine *Phytophthora rubi*, a relative of the potato late blight pathogen, which causes root rot on raspberries. In addition to potatoes, berries (strawberries, blueberries, raspberries, and blackberries) are an economically significant plant in the valley and the Northwest in general. Washington State is one of the largest raspberry producers in the country.

Despite breeding efforts to create disease-resistant varieties, raspberries are susceptible to many root pathogens as well as other aerial pathogens and insect pests. They are very hard to grow organically for those reasons. The wet soil of the Skagit can also exacerbate root rot so that new plantings of raspberries that are supposed to last eight to ten years die out after four or five, a huge expense for growers. In addition to *P. rubi*, there are nematodes, small round worms that live in the soil, that also attack and damage raspberry roots. Although they are not as aggressive as *P. rubi*, they can cause chronic root damage.

My project was to look at alternatives to the main tool used to prevent root problems in soil before raspberries are planted, broad-scale soil fumigation, in which fields are sprayed amply with a pesticide and then usually covered with plastics for weeks to months in order to kill off every living thing below. The chemicals used in fumigation are harsh and include methyl bromide, which is extremely toxic to all soil organisms. Methyl bromide has been banned in Europe since 2005 but is still only slowly being phased out of use in the United States. There are other chemical fumigant options that are less harsh than MB, but they still impact the environment in some way. Several federal grant programs have funded research that explores alternatives to fumigation in many types of crops, from tomatoes to almonds.

In my work, I surveyed commercial raspberry fields for pathogens and looked at alternative ways—which didn't involve synthetic chemicals—to reduce the pathogens. I did greenhouse trials, microplot studies, and field trials. Solarization, the application of plastic over the soil to raise the temperature above survivable levels for most harmful pathogens, was one option. I also looked at brassica seed meals, byproducts of the biofuels industry. These organic materials have in them natural chemicals that can also kill pathogens—without killing every living organism—in the soil, which helps to leave the ground sustainable for raspberry production.

My PhD adviser was a new professor still working toward tenure, a gentle and kind man who did what was asked of him and spent the rest of his time with his family. While I didn't get a lot of mentorship, there wasn't much drama within his program either, and he had plenty of financial and staff support for my project. I could do my work and have my space and that was what I needed. I appreciated his kindness, although I wondered how long he would make it in the fiercely competitive academic world.

Getting back into a research project was fulfilling, but during that time I'd also signed up for extension farm planning classes and was reading everything I could find about farm business planning. Patience didn't come easily, and my fierce attachment to finding my piece of this place was sometimes overwhelming.

Over the months I slowly started to make more friends my age. It seemed that since I had left for Oregon, a lot more people in their late twenties and early thirties had moved to the valley. There was a new art gallery in a small village in the north end of the valley that held monthly shows and openings, and I started to attend their events religiously. The community of old and new artists in the area was inspiring; many artists had ties to a rustic art colony called Fishtown that cropped up in the mid-seventies along the north fork of the Skagit River.

However, I also spent a lot of time alone. I was getting in touch with a new self and the unfamiliar landscapes within that self. After college I had been in back-to-back relationships and never truly on my own, moving continuously through a bizarre array of rentals. Now, most nights, I stayed home with Fran by my side and appreciated how genuinely happy I felt.

I also climbed a mountain. Still unsuccessful in the search for land and nearing the end of my PhD research project, I signed up for a local mountaineering class. It was something to try—and something to distract me from my land search, which had become borderline obsessive. One of the resentments I'd held toward Jim was that he never took me along on his more adventurous climbs. We hiked together, but he saved his technical climbs for his brother and buddies. I would have slowed them down.

The four-month mountaineering course took me into the Cascade Mountains every weekend. Fran stayed with an animal rescue friend and her many dogs, including a three-legged, blind chihuahua. We were a group of around fifteen people, all twenty- or thirty-somethings. In many ways I felt like a young college student again—adventurous and unburdened by responsibility. As we dragged ourselves up snowy, steep slopes (climbers' paths, I learned, are very different from hiking trails), I had never been so thrilled and terrified in my life.

There were several women in the class, three of whom were around my age. One was taking the class to get back into climbing after having two children, another was going through a divorce, and the third was single and settling in the Northwest from the East Coast. We had hours upon hours together every weekend to talk and walk and sweat. A lot! As our numb fingers passed cups of stew and hot chocolate across our campsites, we would often find ourselves giggling like children, enjoying the crisp, fresh cold of the North Cascades in early spring.

The class culminated in summiting Mount Baker. Known to Coast Salish tribes as Kulshan, it's the large active volcano that looms over the Skagit Valley. It took a long day and night to reach the top, but the class had prepared us for it. We had spent one of our previous class weekends dangling in the crevasses on the same mountain. All my worst-case scenarios had run through my head that day as I dangled above the void, above all the realization that I could vanish into oblivion at any moment. (Why had I willfully chosen to sign up for the class?) There was a surge of relief and delight after I finally scurried up and over the icy ridge and back to safety with a successful belay.

On the big day, we spent an evening break at the base camp at six thousand feet; when we woke before dawn we would complete the last four thousand feet to the summit.

In the dark, all four of us were tied together with our class leader, headlamps lighting the way. It was cold, but we knew the sun would be coming soon. There were parties ahead and behind us who were also tethered to each other and loosely attached to the mountain. It was now or never. I stared dutifully at each step I made in the snow, grateful for the chance to be making the climb.

It turned out to be a beautiful sunny day, the perfect weather to make the dream of ascent a reality. At times I fell short of air, and then my legs would start to buckle, but I knew we couldn't stop. We all had confidence in each other. There was a blur of breathlessness and heavy snow and aching calves and then everything flattened out. We put our ice axes down and untethered ourselves from each other and our teacher. At the top we posed for a picture. The sunrise gleaming on our faces made me tear up. I looked out over the entire valley and beyond to Canada. I felt as vibrant as I'd been in the herb fields, and as effervescent as I had ever been. This was the perspective I'd needed, the whole valley sprawled out in front of me. I knew I was exactly where I wanted to be.

⟜

I met Donna, a local dairy goat farmer and real estate agent, through a friend. She lived in the north end of the valley, an area that Henry has always described as having cold and wet soils. Henry was still looking around for land for me through his various contacts, but nothing quite right had come up yet. I hadn't put much attention toward that part of the valley in my search, since properties rarely come up for sale there.

Donna was a tall transplant to the Skagit Valley from Texas. After a career as a labor and delivery nurse, she fell in love with goats and started milking them and making cheese in her kitchen. She started a goat dairy officially when she was fifty years old. While her forward and filter-less Texan personality could be considered aggressive at times, she had a sense of humor too. I had a feeling she could help me find land—my home.

We had three properties on our radar, which she often commented was rare. One was five acres with a big old barn, one was eight acres with two barns and an awkward aboveground pool, and one had a medium-sized barn, a wraparound porch, and the owner's girlfriend still living in the house, stiletto boots and all (there was a confusing divorce happening).

My offer on the first place was immediately rejected.

We were discouraged and anxious. The two-year lease at the rental house was ending and Jim's sister-in-law would soon be returning to live there again. (Maggie had decided to quit her job and travel in Italy and, after the brief house-sitting gig, I would be out of a place to live.) Luckily a friend of a friend offered an extra room in their trailer in the foothills of Blanchard Mountain, also in the north end of the valley. Perched in the middle of a large field, it was basic, but cozy.

At the trailer I had my own room and shared the kitchen. The view of Blanchard Mountain was breathtaking, and as late summer slipped into fall, I'd sit outside and watch the acres of barley easefully sway in the wind. The farm property I'd made an offer on was just down the road, and somewhere in my heart I still hoped to make it mine. But for now, in the rectangular trailer with a large shop out back, I enjoyed being on farmland.

Matthew, my new roommate, loved music and was a skilled crafter who made belts and bracelets and all things leather. He also liked board and video games and often spent long days in his workshop hosting friends who brought what seemed like vats of whiskey. It was a social place, more social than I had been in my days as a student. I started to meet more people through Matthew—and at the local community club down the road. I savored these new connections.

✥

During the first month I lived in the trailer, my grandmother died. Her ten-month battle went fast, and in reflection, it really wasn't a battle at all but an honest truce. When I visited, I would wait by the window for her to return from her treatments at St. Marie's hospital. A troupe of wild turkeys, a few males and several scraggly looking females, had started to descend from the neighboring hills to her front lawn. They would walk up and down the street, as if they were the mayors of the neighborhood on parade, before gathering around the base of her oak tree. I liked to think a wild turkey sighting was good luck.

Somehow, if the wild turkeys were there in the yard, comfort and hope in more years, and visits, to come existed too.

My grandmother was strong and optimistic, but she was also not afraid of death; she accepted it as the natural course of things. In the cemetery she was buried next to Papa and her mother beneath an enormous maple tree. It was autumn, and the maple showed a spectrum from cardinal red to yellow to green. The boldness of color in that tree seemed to hold me up that day.

This was my first experience losing someone really close to me. My grandmother and her mother were both examples of women who'd lived well and fully and without fear. It didn't seem fair to lose either of them this early. I cherished the time I had spent with both of them. I had hoped my grandmother would have many more visits to the valley or at least have seen my prospective farmland. The dream of finding a farm was a dare I could not turn back on now.

◦⊖◦

In those months in the trailer, before moving onto the land that is our farm now, I met Dean. He arrived at one of Matthew's parties with a large box of figs. I recognized him immediately. We had met a few Sundays before at the local bar. His tall, slender frame, rough beard, and gentle, brown eyes were familiar, and I imagined my sister chuckling, "He's your type." He spoke quietly and with intention. He offered me a fig and I couldn't resist. I took another. They were green and soft with sweet purple flesh.

Despite having friends in common we hadn't crossed paths earlier. We both loved the same kind of music and connected over our shared love of the valley. He had also been around the area for many years.

After a few months of living in the trailer, with the beautiful green and golden backdrop of Blanchard Mountain, I was ready for permanence, and I needed to muster up the energy to find another property or find a way to make another offer. The grief over losing my grandmother mixed with my desire for land was overwhelming. One day I had a strong internal nudge to put in a second offer on the farm. I called my mother and she was incredibly supportive, instructing me to set up another meeting with Donna.

I found her at a corner table in a quaint little bakery that sold apple cider doughnuts and was decorated with a glut of anti-shoplifting signs, which always seemed jarring.

"I'm glad we are trying again," I said.

Donna looked at me sternly. "Okay, but I'm not sure what the owners are thinking at this point."

We worked out a competitive second offer, and she pulled out the stack of crisp, white forms to sign and initial. I am not sure what all had transpired with the couple that was selling (I knew several people in the area had been aggressively vying for that place), but they accepted. I would be able to move in by January.

⌁

It was an overcast day, and as a newly graduated PhD, I anxiously paced the entryway of the research center waiting for the director's car to pull into the lot. My cohort of graduate students had heard that Wendell Berry was going to be speaking in Seattle as part of a literary series, and the thought of him traveling to Washington State from Kentucky and not making it an hour north to our vibrant agricultural corridor in the Skagit Valley was tragic. So I wrote to him repeatedly; he responded, and, thankfully, after much discussion (in the form of handwritten letters), he accepted our invitation for a farm tour and lunch with the some of the faculty, students, and local farmers.

Berry had a calm presence, and I was as comfortable around him as if he were a faraway relative whom I'd known through pictures and stories. He sat calmly with his wife, Tanya, in the back seat of our state vehicle, asking questions about what was growing in the fields. Another student was driving, so I got to be the tour guide. Potatoes were going in the ground and many of the bulb crops had been topped over the past month. Red and yellow tulip petals were scattered across the neatly shaped rows. Green and orange tractors were distant speckles.

I was talking more than normal, overcompensating for my nervousness. I wanted to tell him about my choices, my experience in a land-grant college, plans for the future, and the impact of his book, but I stuck to the tour script. His presence was enough.

Throughout the day we also toured some local farms and a bakery and were joined by many of the valley's original organic farmers, longtime devotees of Wendell Berry. In fact, Berry had been the keynote speaker at the first organic farming conference in Washington almost forty years prior. One of the farms that we stopped at that day was a third-generation vegetable farm, and we toured their beautiful tomato greenhouses, ripe with the unforgettable smell of vine and new fruit. It was an honor to be in Berry's presence and to know, in some way, this entire group of dedicated organic farmers.

After we finished our lunch at the research center, I asked Berry to sign my copy of *The Unsettling of America.* It was the first time I had picked up the book since starting graduate school in agriculture.

⊷

The research center, and the graduate program that brought me to this valley, is just a few miles south of our farm, but it feels worlds away now. The average age of farmers is still hanging around sixty, and it is a profession dominated by men. A week before Berry's visit I turned thirty-two. I was getting my footing on the farm, and, according to Henry's rubric, I was finally useful.

Dean is in the kitchen cooking dinner. We aren't officially married yet, but we're living like we are in many ways. He has moved onto the farm full time, letting go of his small cottage twenty minutes south.

Dean is chopping onions professionally. Fingers cupped over the bulb, his knife (the one with his initials scratched into the blade) rises and falls rapidly, but it never leaves the board completely. He has pulled out his black cast iron pan, the one I admired in his kitchen on our first date, the first time he made me dinner. That night I'd noticed he had his own well-used copies of Julia Child's cookbooks.

When we met, Dean had no interest in being a farmer. It was my dream, my farm. He knew the plan was a work in progress. I recall, though, that he was open to the idea of having kids, and maybe a few sheep.

In a former life Dean worked at a bed-and-breakfast in the San Juan Islands. He'd also been a sous chef for many years at a restaurant near Pike Place Market. He'd shared a run-down loft with a friend in South Lake Union, when that part of town was industrial and slightly sketchy—not the Amazon mecca it is today. The loft wasn't technically supposed to be a living space, and it had no heat, but as a young art student he didn't care. It was cheap and he had an electric blanket. Later he left the city and taught himself how to do construction for good money.

The black pan is sizzling now. I smell the onions and garlic. He cooks, and when I return from animal feeding chores in the brisk, fall air, we sit together and enjoy a spicy red bean stew. In my mind, I found this piece of land with the intention of growing vegetables and learning how to raise animals. Of being embedded in a beautiful place and starting a farm-based business of some sort. I wanted a home I could get to know intimately. My plan had been to do it alone. Now it seemed we might do it together—each hard, illuminating season building on the next.

Milk

NURTURING HAS NEVER come easily to me. In the beginning, I would watch the ewes lamb each spring from the back corner of the barn. Occasionally I had to intervene, but for the most part the ewes were strong and independent and made it look easy. In my first season, I remember feeling awestruck by their immediate devotion to their offspring—licking the fluids and pain off the slick, throbbing lambs' bodies the minute they hit the straw.

Since having two daughters of my own, I have found myself apologizing to the friends and family that had babies before me, admitting that I didn't understand the kind of help they needed in the early days and months and years with their new babies. I was oblivious to the energy required to make, and then care for, another human.

What I regret the most is not asking about their birth stories. The gritty details. The bliss that permeates your whole body once the slimy, wriggly babe is placed on your bare chest and arms. The waiting. The pushing. The loss. The joy and release. My two births were like night and day, yet they now blend together in my mind as one long, epic tale, a yarn I am happy telling and retelling, in part because I am proud I survived, and in part because I know my aging body will never give birth again.

Sally has always been a good mother. Standing under the protection of our new sheep barn, I see signs of her imminent labor. This ewe has separated herself from the flock. She is alone against the west wall, alternating between sitting and standing, and she's breathing deeply. When all we had was a small manger, our sheep would lamb only at night, using the protection of darkness to deter patrolling coyotes or hovering eagles. Now, in the safety of four walls and a proper roof, they birth at all hours of day.

It is noon. I have cancelled my dentist appointment. I should know better than to plan anything during lambing season. We try to keep our schedules flexible during the spring. One time, Sally went into labor during a dinner party. Our guests came scrambling out to the barn after sunset and were thrilled to see the new life stumbling around in the hay. Then we all reconvened inside for dessert.

Now Sally is grunting and plunging her top hoof into the straw like a sword. It won't be long until the first lamb arrives. Her top lip curls upward revealing a pale ruddy gum resting above a bottom row of long, angular incisors. I put my hands on her neck and shoulders, letting her know that I am here if she needs me. Her eyes are open, but trance-like. She is here, but not here. I know it won't be long.

As she stands again, pacing around her sacred space (the ewes give each other plenty of room), a fluid bag emerges like a darkened balloon, followed by two small, black hooves and a gleaming white nose encased in the gel-like bag. She scratches the hay until she reaches gravel and then starts to lower to her front, left knee. A final push and sigh and the lamb cascades with a splash onto the barn floor. The body is softly flittering in the moistened hay.

I step in now and use my finger to swipe the mouth, ensuring the airway is open. I make sure to avoid handling the steaming, glassy frame for fear of spreading my scent. A foreign smell can sometimes keep the weary, postpartum mother from claiming a new babe. Within minutes, the lamb is quietly bleating and already trying to stand. Sally begins her diligent cleaning process with loud, lapping fervor. Another is on the way, and probably another, given her girth and birth history. She normally has twins or triplets. A light wind passes through the barn, and the midday sun shines on the slowly regrowing pasture. I always get a surge of adrenalin during the start of a lambing—my call to attention—followed by a calm sense of renewal. All I can do now is give Sally some hot water spiked with molasses and the time and space to finish birthing.

Our first daughter was born on the hottest day of the summer during what had been a long, seemingly unending, drought. She was overdue, and I was a walking mirage, swollen and spilling over with water. I assumed she would slide out easily. However, she became wedged under my pelvis, and no amount of pushing would convince her out. I labored dutifully at home, then at our earth-toned birth center room, and finally (inside the gentle lull of pain medication) in the maternity wing of the hospital.

A surgery team was waiting on the side of the sterile room to swoop in if the doctor couldn't turn her inside of me, if he couldn't use my inner thrusts to pull her out. The original plan, my pre-written birth story, was to have a precious few by my side (my midwife, doula, and husband) to welcome our first girl into the world. Instead, I had a well-equipped army suited up in head-to-toe blue around my bed. Strange, yet caring, eyes peered at me over facemasks that released muffled cheers when my daughter finally appeared without surgery.

Our second daughter, in contrast, shot out of me like a rocket during the Friday night dinner hour. We barely made it to the birth center; my water broke in our van. Alarmingly unmedicated, I faced the difficult task of pushing another human out of my body in a very short period of time. I stumbled into the birthing room like James Brown leaving a stage, while my husband covered my bare shoulders with a robe.

It was fast and fierce, followed by the unearthly relief of head, then shoulders, then plump placenta flopping out onto the bed. I heard one of the midwives whisper, "That was the perfect birth," as I rolled over onto my back to hold my second girl. I had no idea what day or time it was and I was ravenously hungry.

My two birth stories are divergent, two opposing experiences in time and space and ache that have now fused together. I grew up in a family that shared only small pieces of anecdotal (and humorous) information about childbirth with me. Grandma had almost "dropped" her fifth child out on the stairs. My mother recalls the epically long baseball game my dad and uncle were watching while she went into labor with me. When my little sister was born, we were all at home watching *Circus of the Stars* (before I was whisked away in my snowsuit and left in the hospital lobby to sit and sweat and eventually throw up from heat and cigar smoke). No wonder I didn't know what to expect.

Sheep were some of the first animals to be domesticated by humans. However, when they are in labor, they seem wild to me, wholly married to their primal instincts. Even the new mothers just know what to do. I felt so helpless when I was birthing my first daughter. The pain was overwhelming. I grasped for help. I deferred to experts. I didn't know how to listen to my own pounding insides.

Each ewe labors differently. There is not a prescriptive pattern. Each year we learn more as we accumulate more and more birth stories. Some groan while others are as quiet as prayer. Their bodies give them reprieve between contractions, gentle pauses between the primal eruptions of pain— not unlike the rhythm and cadence of human birth.

So far, I have had to assist our ewes on only a few occasions, and my midwifery attempts have been met with bittersweet results. Lambs have died despite my efforts. Lambs have been resuscitated after the trauma of a long, hard labor. Every season holds its own mysterious sorrows and triumphs.

I would like to believe that all of my lambing lessons are making me a better, more attentive mother. Each season I am reminded of how delicate life is and that true caring requires a mix of action, awareness, and compassion. Someday I hope my daughters will want to help me with all the chores. The milking, barn cleaning, shearing, hoof trimming, cheesemaking, and especially the lambing. I want them to see it all. I want them to appreciate the wonder, to witness the fragile boundary between life and death, the arduous beauty of birth. And I will tell them, again and again, how they came into this world.

After years of deliberating about whether or not to build a sheep creamery on our farm, we finally did it seven years after moving in. We had grown organic herbs and produce on the farm for several seasons in addition to having a "few" sheep. Now it was clear that the sheep were not going to go anywhere. My love for them elevated their status from being "just pets" to integral members of our farm.

The sheep were some of my first children, and as I learned to care for them and assist with lambing, I unknowingly cultivated my own maternal instincts. I help the ewes with their births and babes and in return they have made a mother out of me. I realize now that my version of animal husbandry is more like the midwifery support I received during my pregnancies. Having children has made me a better, more attentive, shepherd. I am attuned to their animal bodies, the health of their udders and hoofs, the quality of their milk and wool. The weight and activity of each new lamb is observed and recorded.

Our sheep creamery plans were delayed because of the arrival of our second daughter. She came earlier than expected—like years earlier. We wanted a second child, but we didn't think I would get pregnant again so easily. We embraced our new family timeline, but also realized that constructing a new facility and launching into a milking and cheesemaking routine would not fit into the early months of rearing a newborn. It would be too hard on my almost forty-year-old body—and what was the rush anyway? Starting a creamery was another major life change we were contemplating. So with the birth of our second child, we took an additional season to plan and ruminate along with our ever-growing flock.

As the following lambing season started it was clear that I still wasn't ready despite the year delay, but we had to move forward. We had the plans and grants and loans in place and they wouldn't be there forever. I was severely sleep deprived. Between waking up to feed and sometimes change my daughter, checking on birthing ewes, and making bottles for the orphaned lambs, I was overwhelmed. Along with the late-night chaos, our workspaces (milking parlor, milk house, and creamery) were still unfinished, despite my husband's valiant attempts to complete them before the new lambs arrived. By now we've learned that projects take more time with two small children.

However, the pressure of new lambs dropping nightly forced me to focus on sleep training, and eventually the baby and I found a good rhythm. I got the first ewes on the milking stanchion on April Fools' Day, which that year also fell on Easter Sunday. It was no joke. It was ceremonial, like we had passed a threshold into a new way of life.

I won't lie and say it was an easy routine to adopt. It was a struggle each day to get up early and milk. Day after day. For the first few weeks it was a two-person job, since most of the ewes were not accustomed to being handled and milked. My body was still tired and unbalanced from the energy expended in creating and then feeding a child. My get-up-and-go had got up and left. I reminisced about past (childless) farming seasons when I would wake at 5:00 a.m. to pick squash and make deliveries and was still out weeding at sundown. I'd had the freedom to push myself really hard back then.

There were not a lot of activities I did daily besides brushing my teeth. I'd like to say I did yoga and wrote daily, but that would be a stretch. However, being aware of the many hungry and swelling ewes outside, within earshot, made me get out early to the barn. Before milking the sheep, I fed the baby in our bed. As I woke and felt the ache of fullness in my breasts, I knew that just below my bedroom window, our ewes were ready for that same relief. They were separated from their babies and bursting with milk; it was up to me to be on time. I tended to their needs as much as those of my own children.

I wasn't totally unprepared for the rigid schedule. I'd milked one cow on Gerry's farm and worked a Saturday morning milking shift at Donna's goat dairy during another summer. Previously, we had dabbled in our own homestead milking but nothing on this scale or schedule. Having two small children has brought a lot of regularity to my life: snacks and lunches, naptimes and bedtimes, bath nights and play groups. Our daily activities had already been leaning less toward spontaneity and more toward repetition, making our new project a bit less of a leap than it might have been.

We have had to become efficient too. Even as a small artisanal creamery, we rely on technology and small machines to make our products. Berry writes, "For the true measure of agriculture is not the sophistication of its equipment, the size of its income, or even the statistics of its productivity, but the good health of the land." I'd like to believe that our version of family farming has some real heft to it. I walk in tandem with my small ruminant mothers out to the pastures we planted for them. I pay attention to their health and personalities, how they interact. Our cheese has a story and a flavor that is unique to us. I believe that has to be worth something significant. I believe our type of small farm-based business can be a viable way forward in the overly complicated and exploitative food paradigm we live (and eat) in now.

The poet J. D. McClatchy says that "love is the quality of attention we pay to things." I am striving to raise my daughters well, and be present with them. I am trying to embrace the daily chores and activities this farm requires, even when I am tired and ready to check out. I see my flock and they see me. I feel it in my body—their work and my work, our symbiosis. We have become mothers in tandem, and our work together will continue into the seasons ahead.

Swinomish Reservation, 2011

AS I ENTERED THE NEW campus building, a familiar smell wafted in the front door. Tater tots. There was a small kitchen in front of me and I walked up to the doorway to say hello. Janice, the site manager who had interviewed and hired me, was busily preparing a lunch for all the students. Tater tot casserole. This was something she normally did once a week. It was in her blood to feed and care for others.

"Good morning," she said as I walked through the door. Her long black hair was perfectly straight and gleaming and she was wearing a brown shirt with an ornate silver and turquoise necklace.

"It smells good in here."

"Lunch will be at twelve, right after your first class."

Smiling, I told her I would be there. I had packed a simple lunch, but knew what Janice was making would taste better and be warming on this cold, rainy spring day. Comfort food was always served at this campus, things like tater tots or spaghetti with salad and ranch dressing, which I initially found surprising. What did I expect, salmon and berries? Janice wasn't Swinomish but had grown up on a homestead farm in Montana on the Blackfeet reservation. Her partner was of Coast Salish ancestry and they had two children. Janice had been in the Skagit Valley for many years, having gone to school at a state college before taking a job with the tribal college. She was an excellent caretaker of the site and its students.

Janice, I would learn, also made a huge Thanksgiving feast for the students and faculty every November. All the fixings: turkey, stuffing, and gravy, and multiple pies. Janice's offerings weren't part of any specific meal program; she just did it. The sharing of meals was as important as doing well on tests. That would be my first lesson.

After graduate school at what I thought was our state's only land-grant university, I looked for teaching opportunities. I knew I was staying put in the Skagit Valley and I wanted a break, maybe permanently, from the stress of research. I was under contract for our farm and counting the days. I was imagining what the farm could be and also realizing that it would be a long time, maybe never, until I could farm exclusively for a living.

At a community event I serendipitously learned about the Northwest Indian College, based on the Lummi Reservation outside of Bellingham with several site campuses at various reservations like Swinomish, Tulalip, and Muckleshoot. It was mentioned by one of the tribal elders attending the event, and I immediately went home to look it up. It turned out, I discovered from their website, that the Northwest Indian College is one of our country's thirty-five tribal colleges and universities, designated as land-grant institutions through the Equity in Educational Land-Grant Status Act of 1994, to "help improve the lives and career opportunities for Native students and support research, education, and extension programs that enhance local agriculture and food production."

When I first moved to the Skagit Valley, I lived on the Swinomish Reservation on Indian Road by myself with Jeffrey, my gray cat. I had no idea there were the beginnings of a college nearby. The main campus on the Lummi Reservation had originally started as a school of aquaculture in 1973 before eventually being accredited as a college. The extended campus at Swinomish had started in an office space near the tribal headquarters. The new building was built by the college on rented land nestled within one of the reservation housing developments a year before I started teaching.

I reached out and introduced myself to Janice after I learned about the college. Coincidentally, there was a position open for a new faculty member. The college was looking for someone with a background in gardening and food production. I couldn't believe it! The position would be part time and connected to the Native Environmental Science Program on the main campus. I didn't quite know what "Native environmental science" was, but I was eager to find out.

My relationship with science was confusing as I tried to put all my graduate school experiences behind me and make plans for the farm. Science taught me to see and appreciate minute details in the natural world, from the gentle lobe of a leaf edge to the texture of soil. However, after seven years in a competitive and somewhat solitary work environment, I was experiencing burnout. I had tried to jam myself into the standard mold of a scientist and quite honestly, I didn't fit. The idea of Native environmental science intrigued me. Could there be a different, more holistic way of doing science? One that perhaps also embraces stories and creativity?

In the last few months of graduate school, I had formed a writing group that consisted of some local friends and friends of friends, all women writing in their spare time outside of day jobs like waitressing at the casino or working at the nearby bakery. I was generating more poems than I ever had before in my life. Without consciously knowing it, I was desperately seeking an artistic outlet. After returning from Oregon, I had joined a ceramics class outside of town and started making pottery again, a hobby I had pursued when I was in high school and college but had long forgotten. These circles of women pursuing art, both clay and words, were a lifeline and a mirror reminding me of what my soul loved to do. Now, surrounded by all this new land, I could feel my creativity reawaken.

With many years of school under my belt, I drew on on all my former teachers and class notes, boxed up in the farm basement, to initially design my curriculum, which included basic undergraduate science classes. My education was a wellspring. Also, teaching in this Native environmental science program allowed me a chance to step back and ponder Western science from different perspectives. There were other ways of knowing and appreciating the Earth. As philosopher and professor Greg Cajete writes, "Native science reflects a celebration of renewal. The ultimate aim is not explaining an objectified universe, but rather learning about and understanding responsibilities and relationships and celebrating those that humans establish with the world." Where did knowledge really come from? How can we study land from afar and also feel a close connection to it at the same time? At home as I turned the soil and planted crops on the farm, I realized that, more than just starting a farm, I was beginning a relationship with land. One that carried a lot of responsibilities as well as opportunities for growth. One that might get a little messy at times.

The campus building had a pristine new laboratory classroom and a random assemblage of equipment, like microscopes and wooden plant presses. I was tasked with starting a garden, and I quickly discovered that although there was a lot of overlap between the ethics of organic agriculture—a focus on nature, sustainability, ecological balance—and Indigenous food sovereignty ideologies, I had a lot to learn. Food sovereignty, having a more profound and expansive credo, emphasizes the personal right of the individual to choose and gain access to good food and land. Within tribal communities, food sovereignty is centered around preserving the knowledge of and access to sacred, traditional foods and is inherently linked to tribal sovereignty. Food is personal, spiritual, and essential to identity and place.

Visiting the reservation regularly started to broaden my perspective on the farm and farmland I'd pass on my commute. I wove through familiar potato, berry, tulip, and brassica fields before arriving at Rainbow Bridge over the Swinomish Channel. Although agriculture is the predominant present-day use of this valley's land, all the fields used to be tidelands before the treaties were signed, before settlers built the dike and tide gate system. The channel itself was man-made, dredged in 1937 to create a safe passage for boat traffic. Historically all the land was stewarded, but not owned. As someone coming into "land ownership" for the first time, I realized I didn't really know what that meant and, despite my extensive education, wasn't prepared for the physical, financial, and historical weight.

What native plants were in or around my farm? Where would runoff from my manure and compost go? What types of seeds was I introducing to this ecosystem? How much tillage was too much? How could I work with and take from nature at the same time? And what does sustainability really mean, anyway? The more I attempted to teach Native environmental science, the more questions arose.

The original idea for a garden at this campus came from an informal committee of college faculty, staff, and students; tribal employees; and Swinomish community members interested in creating a living, healing space that supported healthy, wholesome produce and some traditional foods like berries. It dovetailed with other projects in the tribal planning office that included restoring salmon habitat and populations and fostering access to other "first foods" both on and off the reservation, like clams, camas bulbs, and oysters. As I began my position, I was eager to meet with this committee and continue the visioning process.

However, our first meeting was less than successful. No one came. My vegetable soup, made from local leeks, carrots, potatoes, and squash, grown on a farm adjacent to the reservation, went cold. I had sent emails and follow-up phone calls to everyone that I was told might be interested. I thought I had been really polite, but that was not the case. Janice told me to keep holding the meeting space.

"The right people will come," she said.

This would be my second lesson: Be patient and earn trust. Good change doesn't happen quickly.

I spent less time calling and emailing people and started showing up in person to other events and the community acknowledged my presence. After weeks of small and sometimes nonexistent meetings, more people did show up—they could tell I was committed—and eventually we had a group of elders, community members, and students. In the largest of the four classrooms, we would push all the rectangular tables together to make a big square. We shared meals and eventually came up with some plans for the garden. The Swinomish had a long history in the region of cultivating domestic and traditional foods in gardens and wild patches of land for sustenance. I roped Dean into making raised beds.

Every quarter, I met new students of varying ages and academic histories. Some students had grown children and were coming back to school after twenty years to get a degree; others had just graduated from the local public high school. At this campus they had the option of a two- or four-year degree. With small class sizes, a tenet of the school, we were able to take field trips to local nature preserves to observe native plants or sample water quality from the Skagit River. I taught a wide range of classes, from ecology to soil science.

While I didn't have a lot of previous teaching experience, I felt surprisingly comfortable and calm in the classroom. Over time, the students started to reach out and confide in me. The rigidity embodied in graduate school seemed to melt away and I remembered again why I loved science, why I had fallen in love with ecology that first fall I was in college in Vermont. There is so much beauty to appreciate. There is so much mystery to respect. I wanted to give back all I had been given in my education. At home as I turned the soil and planted crops on the farm, I realized that more than just starting a farm, I was beginning a relationship with land. One that carried a lot responsibilities as well as opportunities for growth. One that might get a little messy at times.

⌁

It was a bluebird day in May, and I was walking on the sidewalk that went through the residential village past the campus and down to the waterfront. The reservation housing was a mixture of old and new houses and many extended families shared a single dwelling. I'd often see young children and grandparents sitting outside and playing together. Crab pots and boat motors were lined up along driveways since many men in the village worked as commercial fisherman. Some houses also had firework booths, another common profession, waiting to be set up for the big July sale.

As I stepped off the sidewalk and turned toward the channel that ran through town, I heard the deep thump of a drum coupled with a deep, guttural song. I knew a procession would be starting at the south end of the village. I assumed it would be parade-like; however, this was something else entirely. Dressed in traditional ornate wool blankets and cedar headpieces, the group made its way to the community center with the other spectators—Swinomish families, tribal employees, and their mainly white neighbors—following behind in a long, serpentine line.

The Blessing of the Fleet and First Salmon ceremony is an annual, public event. The tribe invites the entire town to witness this procession and ceremony, which is both a prayer for the safety of fisherman—offered by tribal elders and representatives from the Catholic, Shaker, and Pentacostal churches on the reservation—and an expression of gratitude for the water and all its bounty. Fish are placed in cedar-lined baskets and released in four directions as an offering to the sea at the end of the day.

Before the speeches and songs, the Swinomish cook an amazing meal of salmon, halibut, crab, prawns, mussels, and clams for everyone. The tribal community center and gym is converted into a huge banquet hall. Many of the ceremonies, including winter smokehouse activities, were extremely private, and I never wanted to ask my students, especially the ones who disappeared for weeks of time, too many questions. During important events, like smokehouse and funerals, the school's policy on attendance was extremely flexible. Not everyone subscribed to these traditional practices; multiple congregations and dynamic belief systems existed simultaneously on the reservation, which was hard for me to fully grasp.

However, I felt welcomed at this particular ceremony. A tribal elder, who was also involved with the school and occasionally offered me advice on the garden project, once mentioned that The Blessing of the Fleet hadn't always been so public. The open invitation was intentional, if not a bit political, perennially reminding the community, including mayors and county commissioners, that tribal history and culture are important.

Janice and some of my students had driven down from the campus and were sitting at the end of one of the long tables. I still couldn't believe it was all free. The plates of food were unbearably beautiful, glistening shells and the sunset pink of a well-cooked salmon.

"This is amazing!" I exclaimed to Janice, taking a seat beside her.

She smiled and my students giggled as they looked down at their plates. I was unaware of my own enthusiasm.

"There must be like five hundred people here," I said noticing the line that still curved out the door.

One of my students looked up at me with a diffident grin.

"We like to say that when the tide is out dinner is served."

I hadn't heard that yet, but would hear it many more times until it finally sank in, the direct connection between traditional foods and spirit.

⌖

Driving home that day, past the rolling hills and newly planted potato fields, I thought about how far back that ceremony must go. How well these families know these waters and shores; it's in their DNA. How intertwined food and place can be. A deep sense of grief, guilt, and hope swelled up within me, wondering if we will all eventually figure out how to live here together.

The Swinomish have an interesting relationship with science. They are the first tribe in the nation to have a climate change action plan and they employ an army of scientists and lawyers at the ready to monitor water quality and salmon habitat issues throughout the region and bring them to the state's attention. There are long-seeded political battles with farmers and other stakeholders over how to manage land and waterways, and the Swinomish have been excellent leaders in climate adaptation and environmental policy across the state. While adhering to traditional knowledge and practices, the tribe generates some of the most progressive ecological science in the country and pushes land management strategies toward greater ecological balance, using their own land as a testing site.

To expand the garden, buy more supplies, and offer more community classes, our committee was able to get some grant funding from a national food sovereignty organization. One of the expectations of the grant was that we would attend and participate in food sovereignty conferences to share our work.

One of the national conferences was hosted by the Oneida Tribe in Green Bay, Wisconsin. This event itself looked fascinating; I was also excited to visit my homeland. I couldn't wait to return to this beloved region where my parents grew up, where I was born, where my grandmother had lived and died, and where my father's parents were still living at the time. However, as I arrived at the casino conference center, a place my family drove by every summer we had flown out there to visit relatives, I was embarrassed by how much I didn't know about the Indigenous history of this place. My grandfather played golf at the Oneida Golf Club, and I never questioned or asked about the origins of that name.

Janice had traveled with me to a food sovereignty conference in Phoenix, but this time I was alone. My aunt Dee was still living in my grandmother's house a few towns away, so I was able to stay with her. The conference program was fairly packed, so we didn't have a lot of time together, but it was still so comforting to be around her and all my grandmother's belongings. Dee had made the house her own, but I was grateful for my grandmother's presence as well.

The Oneida tribe were wonderful hosts, warm and inclusive. Tribal members from all over the country had come to share their food projects and make connections. In the trade show I met a man of Anishinaabe heritage who was selling maple syrup. I had never even considered maple syrup to be a traditional food, but of course settlers' knowledge of when and how to tap those trees had to have come from somewhere. Now, he told me, the maple syrup industry is a complex corporate hierarchy and most tribal members are disconnected from not only the profits of their traditional technology, but the trees themselves. Collecting syrup used to be a seasonal, family endeavor for tribes like the Ojibwa, Cree, and Algonquin. Sadly, I heard the same story of biocolonialism repeated as I moved down the line of tables talking to people about their work reclaiming traditional foods, such as wild rice and corn.

On one of the final days the Oneida offered a tour of their integrated food system. We visited the Oneida tribe's nearby organic farm, where they were learning to recultivate a traditional white corn variety, one that has 18 percent protein compared to 5 percent in contemporary sweet corn. In a state dominated by genetically modified feed corn and soybean rotations, this was not easy work, but they were trying. We also got to see their buffalo ranch and appreciate these magnificent animals. They ethically raise meat and sell it for profit. Finally, we got a tour of their forty-acre apple orchard and processing facility, where cider is made and sold in a small retail shop.

In terms of a business model, it was all very impressive, and the Oneida tribe's commitment to place and food, combining old and new foods, was inspiring. On the final evening of the conference, various tribal youth from the area performed dances and songs for the attendees, dressed in traditional clothing. Under the fluorescent glare of the conference ballroom, the performers still glowed with pride. I didn't understand the language, but I could feel the power of each sound, the intentionality of it. It was a privilege to be in that room that night, to accept this offering. Despite my long history with this area, my fond family memories, the history was much more problematic than I ever knew.

⊕

Back on campus, four of my students, all women, who were going to graduate at the same time had formed a cohort. They had been taking classes on and off for several years, but were now close to finishing the four-year degree program. One student was in her early twenties, another in her early thirties, and two were returning to school after raising their families. It was a close-knit group and I admired how they all helped and supported each other. Age didn't matter.

As part of one of our final classes together, we took a field trip to Neah Bay. After a long and winding drive in the white, twelve-passenger van I rented, we ended up at the coastal Makah village. Janice came along with the class and brought her daughter, who was now a student in the college. Everyone had gotten a little carsick from the last leg of the drive, but we were all happy to be here. We were at the end of the world.

We walked the massive U-shaped beach outside our cabins on Makah Bay. Light danced off the frigid waves and broke in layers. I didn't have the words to describe what I was feeling. Something like home, something like contentment. It was good to be away, but at the same time there was a familiarity here. The smells and sounds on this tip of the Olympic Peninsula are the same as the ones on our northern beaches around Fidalgo Island and out in the San Juan Islands, places I have explored since I was a child, only here they seemed larger in magnitude and scale. I looked at my students, realizing this has always been home for them and many, many, many ancestors before.

The next day we visited the Makah Museum, filled with artifacts recovered from a nearby Makah village site as well as full-sized replicas of canoes and a longhouse. Archeologists determined that there was a large earthquake around five hundred years ago, which has been corroborated by oral histories within the Makah, that shook a water-saturated hill above the village of Ozette. It buried several longhouses, but the wet, oxygen-free clay preserved much of the site. After eleven years of excavation, researchers and tribal members had retrieved over fifty-five thousand artifacts, many of which are on display at the museum.

We spent many hours at the museum; for several students it was their first time here. While they all told me they found it interesting and help-ful, I wondered what it felt like to see a way of life exhibited like this (the Swinomish did not have a museum on their reservation). A way of life that was both historic and modern, actively practiced and also archived. I could never see the world through their eyes. It was also clear that I would always feel like a visitor—at the college, on my farm, in Washington and its uneco-logical borders, even in this young country where I was born and raised.

✧

By the time I was pregnant with my second daughter, my time at the tribal college was coming to an end, but I would never see things in the valley the same again. While I still admired the farmers around me, I carried some new and important truths from my time at the college. Nothing was more important than taking care of where you live and the people in your life, and the patriarchal way we try to make nature perform and produce for us will never succeed. There has always been a better way.

It was time for me to move on, the call of early motherhood being both oppressive and irresistible. Now, however, moving on did not equate phys-ically moving. I wasn't going anywhere different geographically. Most of my students from the tribal college weren't planning to move away to the next big opportunity, whether that be a job or graduate school, which ini-tially surprised me. They wanted to grow and learn and stay. Sure, some did leave for graduate school or to travel, but there was always a return. It occurred to me that being a place-based person is not necessarily a choice.

That would be a third lesson: there is freedom in restraint; life can in-deed be a circle.

Moon

THE FULL MOON RISES over the blackberry bramble along the ditch. It has been shining so bright these past few nights, aiming light into all the dark spaces, memories, regrets. I pay attention to moon cycles now, more than before, the waning and waxing, the pull of tides and push of emotions that comes with each phase, the odd behavior of animals under the full moon.

Berry writes, "Ways of life change only in living."

Below the moonlit driveway I see the tall outline of our old barn. The one built in the 1930s. The one that has been saved: the posts reset, the south siding replaced with clear vinyl panels. The previous owner used part of barn as a well-lit workspace. He built a boat there and sailed away on it. It has been years since this place was farmed in any real sense. The new barn to the north—the one we built—glimmers under the lunar glow, light reflecting small miracles off the solar panels.

I could read a book out there in all that brightness, but I retreat inside, assured that the next day will come soon enough. And I need rest. It's summer and it's already been a long season. It's summer and I am counting the days until fall. The absence of choking wildfire smoke this August is a small comfort. The sky is clear tonight and rejuvenating.

Since we milk the sheep every day, it's hard to get away this time of year. In the fall, when they are dried off and bred for the next season, we get a break. For now, and for the foreseeable seasons to come, summer will always mean go. Heat equals work and motion—and long days. I keep reminding myself that I can relax and visit with friends and family in a few short months. We might even take a trip in October. For now, I must surrender to the activity. The constant needs and little fires. A sick sheep, a last-minute order. The broken cooler, full of freshly made cheese that must stay below 40 degrees Fahrenheit or we are legally obligated to throw it all out. The wilting plants begging for water. We're giving a farm tour to a school group next Tuesday. It's nonstop.

But it's not all chaos. There is a rhythm in what we do. I feel proud working this hard. I am trapped, but I am free. Our small sliver of tideland-turned-delta is so beautiful I don't want to go anywhere else anyway, at least not at this time of year.

⌁

Dean is out back tilling in a few rows of purslane. We have been harvesting from those plants for weeks, and now they have bolted to swollen, unmarketable, yellow flowering heads. The rich, fleshy plant grows here as a weed, but it's becoming more popular in restaurants, so we grow an upright variety from seed. It has a lemony, sweet taste and was traditionally a staple ingredient in Lebanese and Mexican cooking. We seed it three different times in the growing season to make sure we always have new, succulent plants. Purslane goes to seed when the temperature gets unbearable.

I am inside with our two girls. We can't afford a full-time babysitter, so in the afternoons, my husband and I alternate, fusing into one person running the farm. During this particular day I have milked the sheep and made a batch of cheese. Now I am coloring and eating carrot sticks and apple wedges while he does the tractor work. He will feed all the animals and come in to make dinner. While he makes dinner, I will go out to finish chores. An agrarian merry-go-round.

The sheep will be rowdy because of the moon. They will make zigzag jaunts around the barnyard. I'll find it amusing and then, realizing how tired I am, annoying.

We will eat together out in the late evening light on the porch, with fresh blackberries for dessert. I am a good cook, but he is a great cook. It's how he shows love. His portions are generous. Each ingredient, each flavor, is deliberate. I make a mess; he cooks and cleans methodically. He applies that same care to the girl's lunches, cutting their sandwiches in perfect squares, refining his fruit leather recipes, or whipping up a dish of home-made mac 'n' cheese.

It's my turn to do bedtime. I will wash my hands and change my pants before taking the girls upstairs to read stories and sing songs. Right now, "The Farmer in the Dell" is a favorite. I laugh as the song begins, since I've changed the lyrics from "The farmer takes a wife" to "The farmer loves her life." They don't notice the difference yet, but they will in time.

It's still remarkable to me that two girls came from my one body. They are twenty-two months apart, and because of either our apparent lack of planning or my exceptional fertility, they were both born in the farm season. Even though I waited until my late thirties to have children, I fortunately did not have any issues getting pregnant. A few tries and the deal was done, which I know wasn't the case for a lot of my friends. I don't take this for granted, even though our daughters both were born at inconvenient times.

Eloise, our youngest, was born in May, and June, now four, was born in July. I didn't realize how hard we had been working on the farm until we added new babies to our summer work. The last few seasons have been a blur. And now these two blue-eyed enigmas (my husband and I both have brown eyes) are staring back at me waiting for another verse.

It hasn't quite sunk in that this is my life now. That I am their mom.

My youngest daughter is hanging on my shoulders now, and the oldest dances around the room to the new song I've started singing. "Jingle Bells." I sing it year-round because they love it and I love it and we all know the words. We defy holiday music rules and all sing together. I want my girls to be outspoken and empowered and uninhibited. I also want them to love holidays, because I do.

"Again, Mommy!" they both shriek in unison. Time isn't a concept they can grasp yet and I remind them that it is almost time to fall asleep. A few more rounds and they will both settle into their beds. Eloise, unconsciously, shows she is tired by putting both her hands behind her head. June twirls her hair around her first finger. They both get a soft glaze before fading into sleep. The blinds are closed so they can't see the brightness of the lingering moon. They don't know the grass outside is glowing. They also don't realize that we had to bury a lamb today. There is a lot you don't tell toddlers.

Right now it feels good to lie on the thick bedroom rug and listen to their perfect breathing. Echoes of a baseball game rise from the living room. I'll zone out here for a little bit longer before going downstairs to pour a glass of wine and sit out on the deck. There is a slight chill in the air—a sensation that saddens and excites me. In the morning we will rise with the sun and do it all again. Kid routines and animal routines. Feeding and cooking, eating and making, define our days.

I moved onto the farm property almost ten years ago, and I will never forget the crazy explosion of spring. The land transformed and I wasn't prepared. The pastures, fruit trees, and front lawn all became small jungles and I didn't even own a mower, except for a small orange electric contraption a friend had given me for free.

It's taken several seasons of watching things grow and die here to develop a practical management plan. Now we intensively rotate the sheep through the pasture, moving them after they have eaten the grass down to no less than three inches. We use landscape cloth in our commercial herbs to block the weeds. Another friend, who owns a raspberry farm, prunes the fruit trees in the winter. We start ripping the morning glory out of the garden rhubarb in June, but it always comes back, every year.

I guess if it didn't return, I would be upset. Our rhythms would feel off. We can't cut corners; all the jobs need to be done—and at the right time of year—so we can feel one step ahead of the constant, fierce push of nature, its urge to grow and spread.

The other day I saw a news story featuring a fishing village in China that had been abandoned and was now being "returned to nature." All the buildings were green, colonized by opportunistic plant life. I know that feeling well. The steady creep of chickweed between newly transplanted herbs. The proliferation of thistles in an otherwise suitable paddock.

Farming, no matter how natural, has always been about resistance to nature in one way or another. Altering the landscape for gain, making space and order within its wild brilliance.

It is me and the moon again, outside. The cool calm on the deck offers solace. The farm is a destination I've reached and, still, a blank slate in many ways. I am writing this story anew each day. The girls are definitely asleep now, and Dean is half asleep in front of his baseball game. I hear a barn owl screech and imagine that the moonlight makes hunting easier for her. I hear the constant, distant din of the neighboring farm's irrigation pump. A low moan, a summer siren.

Skagit Valley, 2019

PASTURE HAS its own grace.

⟡

When you look out on the late August light reflecting off the rich, verdant mix of grass and clover, you are looking at the place where you hope to be forever. You didn't grow up on this farm, but right now you can't imagine living anywhere else. You own land, but you know it is not really yours.

To ruminate means to think deeply about something. It also means to chew cud. Sheep are small ruminants. Instead of a one-compartment stomach, they have four. The rumen, the largest of the four compartments, is the digestive center filled with billions of tiny microorganisms that are able to completely break down forage. Ruminants partially chew a large amount of forage each day. The rumen acts as a large fermentation vat, completely breaking down all they have eaten, thus increasing the digestion rate.

To you, it seems like science and magic.

In psychological terms, rumination is the focused attention on the symptoms of one's distress and its possible causes and consequences. As you stare out on this open field you feel, first, gratitude for being here and, second, remorse that farms and farmers in this country are still disappearing at an alarming rate. This is due to a rash of reasons, including globalization and consolidation. You have known about these things for a long time, and as much as you have put your heart into regenerative food production over the past two decades, you feel small. Maybe too small to make a real difference. You are not sure what the future will hold.

⤙

"The modern mind longs for the future as the medieval mind longed for Heaven. The great aim of modern life has been to improve the future—or even just to reach the future, assuming that the future will inevitably be 'better,'" writes Berry.

⤙

Eagles watch us. They talk to each other from the tops of cedar trees, and a mating pair is now making a nest in the old poplar tree in the back corner of the farm. It's a blessing to have their nest on our property. A sign, I hope, that we are doing something right, producing both food and refuge within one farm ecosystem.

I believe that is part of our obligation. Yes, we are trying to run a farm, but the land itself is a stakeholder that needs to be treated well. Money and ecological balance are not necessarily bedfellows. The planted pasture will continually feed our sheep and in turn hopefully feed us and restore soil health. It is a long-term investment with mutualistic benefits, linking us indefinitely with the land and the integrity of our animals.

This valley, with its curvy rivers and luscious sunsets, took hold of me a long time ago. I was supposed to just be a visitor. A student, doing research. Temporary. Passing through between experiences, like so many of the tourists that jet along Highway 20 to catch a ferry to the San Juan Islands to the west or to hike the Cascade Mountains to the east. But I decided to stay. And staying never came easily to me.

I am grateful for my education, especially the many years I spent both learning and teaching at land-grant colleges. However, after being on the farm and learning from the land itself, I can't help but think that the initial land-grants institutions (founded in 1862 and 1890) in particular have lost their way. They have strayed from their core mission: to educate and generate practitioners, to champion farming as a profession. Instead, they are living out the criticisms and prophesies of Berry that are now as old as me. Specialists run the show, and there is little room for interdisciplinary exploration and empathy within the narrow confines of applied Western science, especially when the work is largely funded by profit-oriented corporations.

⤴

My heart, it turns out, leans toward poetry.

⤴

In learning how to farm, I have learned how to let go of certain freedoms. The farm comes first, always. I can't help but think about climate change, and what it will take to make the much-needed household and community-wide changes in lifestyle, the ones we have all needed to make over the past two decades. The ones I have been trying to understand and embody since I was in college. Fewer trips, eating locally and in season to reduce food miles, and paying farm workers fairly—which means paying more for food. Needing and wanting less in general.

Even as a farmer I will admit that I have too much stuff, use too much plastic, both on the farm and in our home, and I seek out a lot of modern conveniences. I am not a Luddite, and I don't pretend to be. When I look back to the old models for homesteading and living off the land, I see the exploitation of women and minority labor at the forefront. We need new models of land management that transcend gender, race, and class.

Hopefully, we can serve as one type of example of how to live and farm sustainably, but Dean and I are far from perfect. I am not here to preach.

Out of the blue one day I got a call from a *New York Times* reporter. He was asking people from different industries to comment on the future. I remember talking to him as I bounced on an oversized yoga ball, my young baby girl finally asleep in my lap. "Sometimes, it's scary looking forward as a farmer. From our farm, we can see Mount Baker and the Puget Sound, a volcano and a rising sea. We're kind of living for the moment, in a geological sense."

⤴

I still know this to be true.

⊹

In writing about this life, on this farm, I have come to realize that this little bit of land allowed me the space to dig into deep wounds and my own self-doubt. The intense physical activity of farming stimulates the senses and makes you keenly aware of where you are in space and time. You have to be paying attention, you have to work at nature's pace, you have to submit to the cycles of energy and action and lose the idea that you are in control.

⊹

Failure lurks around every corner: insect-ridden plants, a sick animal, the perpetually rising seed and feed costs.

⊹

The sheep head into the barn for the night, single file. Tomorrow they will be milked and set out to pasture to do it all over again. In the winter months, they will be fed hay and grain and kept near the barn so the pasture can rest. Pasture to rumen, rumen to milk, milk to cheese.

⊹

Over the years I have written to Wendell Berry a few times and, thankfully, he's responded. I am not sure if he finds my letters interesting or not, but he listens—to my plans, my ideas, my worries. Recently I wrote to him again and told him our sheep creamery and Grade A dairy became licensed in 2018. That we had a second daughter in 2017. I was finally writing more, working on a memoir. I wrote that I was happy about what we were doing, but that I was also really tired.

His response arrived quickly and was brief, just a few lines. One line stood out in particular. "Don't take on so much work that you overwhelm your gratitude."

Initially, I was furious, my cheeks and neck feeling hot.

Then, after I calmed down, I blushed in embarrassment. Dear God, how could I be mad at this person who is so important to me? This very prominent person who bothers reading my letters and responding in his own handwriting! The agrarian prophet, whose words have inspired me for over two decades!

What did I want him to say anyway? Maybe I had hoped for some meaningful, fatherly praise for my efforts—or for him to tell me that this was the "right" thing to be doing with my life and that I was, in fact, having an impact on the world. After reflecting, and noticing how much the last few years—two girls and a new creamery—had taken out of me, I knew he was right. He wasn't saying don't be tired, don't try . . . just, be careful.

There is a lot about the future that seems unpredictable to me and farming itself is no exception. Small farming is almost all risk. A wet spring or an injury could change everything, and I confess that at some point my aging body might give in to the urges to read and write daily from the comfort of my office year-round—a routine now confined to only a few winter months. It is a privilege to have choice. But for the time being, while I still feel strong and motivated, I want to do this work of tending healthy soil and animals.

Nestled within my usefulness there is a sense of belonging. I am trying to observe as much as I can, and I'm taking good notes. We are caring for this landscape as best we can. We are raising strong women and sharing good food. It is all I've ever wanted.

Acknowledgments

I AM GRATEFUL to the many people who have supported me in starting and, eventually, finishing this book that has undergone many forms and titles over the past many years. Some unsuccessful poems morphed into essays that eventually blossomed into a memoir. Thank you to Ellie Bryant, Anna Ferdinand, Jill Swenson, Christianne Squires, Suzanne Morrison, and Kary Wayson for their careful edits along the way. My 2019–2020 Hugo House Book Lab Cohort (Minda Lane, Vera Giles, Jon Strickland, Lauren Allen, and Katherine Shaw) led by Claire Dederer was invaluable not only in shaping the final draft, but in keeping me motivated and engaged in this work during difficult times. Claire, thank you for your crucial comments. Memoir classes with Simone Gorrindo and Elissa Altman were also very helpful as I timidly stepped away from poetry and ventured into the often-intimidating land of the personal essay and memoir. Larry Campbell served as a sensitivity reader for "Swinomish Reservation, 2011." Thank you, Larry, for your time, feedback, and wisdom.

Thank you to Oregon State University Press and your amazing team for bringing this book to life and believing in me and my work.

I value every person and institution mentioned in *A Little Bit of Land*— they are part of my story and teachers in one way or another, and I am grateful I had a chance to learn from them. Thank you to my husband,

Dean, and my mother, Mary, for reading and listening to this book as a work in progress and for offering time and childcare to let me finish this project. June and Eloise, love you always! Final edits were done in part at poet Susan Rich's fabulous Alki Beach Writing Retreat.

Previous versions of certain chapters appear in *Taproot, Comestible,* and *Gastronomica,* and the 2020 Sacred Essay Contest for the Center for Interfaith Relations (Finalist "Resettling of the Mind"). Thank you Rosemary Winslow and Catherine Lee for including "Birth" in the anthology *Deep Beauty: Experiencing Wonder When the World Is On Fire.*

To the authors quoted in this book that inspired me to write—namely Wendell Berry, Robin Wall Kimmerer, Maxine Kumin, Seamus Heaney, and Emily Dickinson—I am your forever fan.